From Natural Polyphenols to Synthetic Bioactive Analogues

From Natural Polyphenols to Synthetic Bioactive Analogues

Editor

Corrado Tringali

MDPI • Basel • Beijing • Wuhan • Barcelona • Belgrade • Manchester • Tokyo • Cluj • Tianjin

Editor
Corrado Tringali
Università di Catania
Italy

Editorial Office
MDPI
St. Alban-Anlage 66
4052 Basel, Switzerland

This is a reprint of articles from the Special Issue published online in the open access journal *Molecules* (ISSN 1420-3049) (available at: https://www.mdpi.com/journal/molecules/special_issues/polyphenols_synthetic_analogues).

For citation purposes, cite each article independently as indicated on the article page online and as indicated below:

LastName, A.A.; LastName, B.B.; LastName, C.C. Article Title. *Journal Name* **Year**, *Article Number*, Page Range.

ISBN 978-3-03936-704-7 (Hbk)
ISBN 978-3-03936-705-4 (PDF)

© 2020 by the authors. Articles in this book are Open Access and distributed under the Creative Commons Attribution (CC BY) license, which allows users to download, copy and build upon published articles, as long as the author and publisher are properly credited, which ensures maximum dissemination and a wider impact of our publications.

The book as a whole is distributed by MDPI under the terms and conditions of the Creative Commons license CC BY-NC-ND.

Contents

About the Editor vii

Corrado Tringali
Special Issue: From Natural Polyphenols to Synthetic Bioactive Analogues
Reprinted from: *Molecules* **2020**, *25*, 2772, doi:10.3390/molecules25122772 1

Nunzio Cardullo, Vincenza Barresi, Vera Muccilli, Giorgia Spampinato, Morgana D'Amico, Daniele Filippo Condorelli and Corrado Tringali
Synthesis of Bisphenol Neolignans Inspired by Honokiol as Antiproliferative Agents
Reprinted from: *Molecules* **2020**, *25*, 733, doi:10.3390/molecules25030733 5

Denise Galante, Luca Banfi, Giulia Baruzzo, Andrea Basso, Cristina D'Arrigo, Dario Lunaccio, Lisa Moni, Renata Riva and Chiara Lambruschini
Multicomponent Synthesis of Polyphenols and Their In Vitro Evaluation as Potential
β-Amyloid Aggregation Inhibitors
Reprinted from: *Molecules* **2019**, *24*, 2636, doi:10.3390/molecules24142636 23

Xiaopu Ren, Yingjie Bao, Yuxia Zhu, Shixin Liu, Zengqi Peng, Yawei Zhang and Guanghong Zhou
Isorhamnetin, Hispidulin, and Cirsimaritin Identified in *Tamarix ramosissima* Barks from
Southern Xinjiang and Their Antioxidant and Antimicrobial Activities
Reprinted from: *Molecules* **2019**, *24*, 390, doi:10.3390/molecules24030390 43

Ferdaous Albouchi, Rosanna Avola, Gianluigi Maria Lo Dico, Vittorio Calabrese, Adriana Carol Eleonora Graziano, Manef Abderrabba and Venera Cardile
Melaleuca styphelioides Sm. Polyphenols Modulate Interferon Gamma/Histamine-Induced
Inflammation in Human NCTC 2544 Keratinocytes
Reprinted from: *Molecules* **2018**, *23*, 2526, doi:10.3390/molecules23102526 59

Gérard Lizard, Norbert Latruffe and Dominique Vervandier-Fasseur
Aza- and Azo-Stilbenes: Bio-Isosteric Analogs of Resveratrol
Reprinted from: *Molecules* **2020**, *25*, 605, doi:10.3390/molecules25030605 75

About the Editor

Corrado Tringali is a Full Professor in Organic Chemistry with the University of Catania, Italy. At present, he is a Member of the Board of the International Doctorate in Chemistry and has been President of the Master of Science in Chemical Sciences at the University of Catania, Member of the Scientific Committee of the International Summer School on Natural Products "Luigi Minale" and "Ernesto Fattorusso", and Chairman of the International Doctorate in Chemistry at the University of Catania (academic years 2006–2011). Prof. Tringali is a senior researcher in Chemistry of Natural Products. He was trained in bioassay-guided purification methods at the Ècole de Pharmacie (Lausanne University) as well as in modern NMR techniques at the Institute für Organische Chemie und Biochemie (Bonn University). His recent research activity has focused on isolation, characterization, and synthesis of bioactive polyphenols. He is author or co-author of 140 publications in international journals, including reviews and chapters for some books. He was invited by Taylor & Francis–CRC Press publishers as the Editor (and co-author) of the book Bioactive Compounds from Natural Sources: Isolation, Structure Determination and Biological Properties (2001), The Second Edition, with the subtitle Natural Products as Lead Compounds in Drug Discovery, which was published in 2012.

Editorial

Special Issue: From Natural Polyphenols to Synthetic Bioactive Analogues

Corrado Tringali

Department of Chemical Sciences, University of Catania, Viale A. Doria 6, 95125 Catania, Italy; ctringali@unict.it

Received: 12 June 2020; Accepted: 12 June 2020; Published: 16 June 2020

In recent years, phenolic compounds from plant sources, commonly referred to as 'plant polyphenols', have been the subject of an impressive number of research studies, to a large extent focused on the healthy properties attributed to diet polyphenols, including antioxidant, anti-inflammatory, antineoplastic, antidiabetic, neuroprotective, and other biological activities. Additionally, phenolic compounds isolated from toxic plants and showing cytotoxic or antiproliferative activity have been intensively investigated in view of a possible exploitation of their anticancer properties. In parallel, many research groups have focused their work on obtaining synthetic or semisynthetic analogues of these molecules, with the aim of enhancing their biological activity and possibly improving their metabolic stability and bioavailability, as a first step towards the discovery of new chemotherapeutics agents. The preparation of libraries of analogues derived from natural polyphenols may also contribute to a better understanding of the molecular mechanisms of action of the most promising compounds through structure–activity relationship (SAR) studies. Finally, synthetic compounds inspired by a natural scaffold may also show new and unexpected biological properties.

Thus, this Special Issue aims to highlight recent results both in the field of natural polyphenols and in that of their synthetic bioactive analogues. It is composed of one review and four original articles, overall reporting results about the synthesis of antiproliferative bisphenol neolignans inspired by honokiol, a multicomponent synthesis of polyphenols as potential β-amyloid aggregation inhibitors, polyphenols from *Tamarix ramosissima* and *Melanoleuca styphelioides* as potential antioxidant, antimicrobial or anti-inflammatory agents, and a review article on aza- and azo-stilbenes as bioisosteric analogs of resveratrol.

Cardullo et al. [1] report the synthesis of a library of bisphenol neolignans inspired by honokiol, a natural polyphenol showing a variety of biological properties, including antitumor activity. The natural lead was subjected to simple chemical modifications to obtain a first group of derivatives. To obtain further neolignans with a different substitution pattern to honokiol, the Suzuki–Miyaura reaction was employed. These compounds and the natural lead were subjected to antiproliferative assay towards HCT-116, HT-29, and PC3 tumor cell lines. Six neolignans show GI_{50} values lower than those of honokiol towards all cell lines. Three compounds showed GI_{50} in the range of 3.6–19.1 µM, in some cases lower than those of the anticancer drug 5-fluorouracil. Flow cytometry experiments showed that the antiproliferative activity is mainly due to an apoptotic process.

The paper by Galante et al. [2] describes an example of application of the Ugi multicomponent reaction to the combinatorial assembly of artificial, yet "natural-like", polyphenols. The authors used a "natural fragment-based approach" to the combinatorial synthesis of polyphenolic molecules. Starting from small phenolic building blocks, they obtained a series of artificial polyphenols, which were evaluated as inhibitors of β-amyloid protein aggregation and potential anti-Alzheimer agents The biochemical assays highlighted the importance of the key pharmacophores in the synthesized compounds. As final result, a lead for inhibition of aggregation of truncated protein AβpE3-42 was selected.

A further contribution by Ren et al. [3] is focused on polyphenols from *Tamarix ramosissima* bark, to determine their potential antioxidant and antimicrobial activities. A total of 13 polyphenolic compounds were identified by UPLC-MS analysis. Hispidulin and cirsimaritin, active ingredients of traditional Chinese herbs, were identified for the first time in a *Tamarix* sp. The main constituents of bark extract are isorhamnetin (36.91 µg/mg extract), hispidulin (28.79 µg/mg) and cirsimaritin (13.35 µg/mg). The antioxidant activity of the bark extract was evaluated through DPPH, ABTS, the superoxide anion and hydroxy radical scavenging, ferric reducing power and FRAP. Promising results were obtained for DPPH (IC_{50} value of 117.05 µg/mL), hydroxyl radical scavenging (151.57 µg/mL) and reducing power (EC_{50} of 93.77 µg/mL). The *T. ramosissima* bark extract showed antibacterial activity against foodborne pathogens. *Listeria monocytogenes* was the most sensitive microorganism with the lowest minimum inhibitory concentration (MIC) value of 5 mg/mL and minimum bactericidal concentration (MBC) value of 10 mg/mL, followed by *Shigella castellani* and *Staphylococcus aureus* among the tested bacteria.

Albouchi et al. [4] present a study on *Melaleuca styphelioides*, known as the prickly-leaf tea tree. The authors characterized the polyphenols extracted from the leaves and determined their potential antioxidant and anti-inflammatory activity. LC/MS-MS was used to identify and quantify the phenolic compounds. An assessment of the radical scavenging activity of all extracts was performed using DPPH, $ABTS^+$ and FRAP assays. The anti-inflammatory activity was determined on interferon gamma (IFN-γ)/histamine (H)-stimulated human NCTC 2544 keratinocytes by Western blot and RT-PCR. The methanolic extract presented the highest concentration of phenolics. The main constituents were quercetin, gallic acid and ellagic acid. DPPH, $ABTS^+$, and FRAP assays showed that methanolic extract exhibits strong concentration-dependent antioxidant activity. IFN-γ/H treatment of human NCTC 2544 keratinocytes induced the secretion of high levels of the pro-inflammatory mediator inter-cellular adhesion molecule-1 (ICAM-1), nitric oxide synthase (iNOS), cyclooxygenase-2 (COX-2), and nuclear factor kappa B (NF-κB), which were inhibited by the extract. In conclusion, the extract of *Melaleuca styphelioides* can be proposed as a useful treatment for inflammatory skin diseases.

Finally, this Special Issue includes a review article by Lizard et al. [5], devoted to aza- and azo-stilbenes as bioisosteric analogs of resveratrol. Stilbenoid polyphenols are well known for their promising biological properties. However, their moderate bio-availabilities, especially for trans-resveratrol, prompted a number of researchers to optimize their properties by synthesizing innovative resveratrol analogs. The review is focused on isosteric resveratrol analogs, namely aza-stilbenes and azo-stilbenes, in which the central double bond is replaced with C=N or N=N bonds, respectively. The biological activities of some of these molecules are reported in view of their potential therapeutic applications. In some cases, structure–activity relationships are discussed.

We expect that this Special Issue will promote interest in the search for bioactive polyphenols as potential therapeutic agents.

Funding: This research received funding from 'Piano della Ricerca di Ateneo 2016–2018, Linea d'intervento 2' of Università degli Studi di Catania.

References

1. Cardullo, N.; Barresi, V.; Muccilli, V.; Spampinato, G.; D'Amico, M.; Condorelli, D.F.; Tringali, C. Synthesis of Bisphenol Neolignans Inspired by Honokiol as Antiproliferative Agents. *Molecules* **2020**, *25*, 733. [CrossRef] [PubMed]
2. Galante, D.; Banfi, L.; Baruzzo, G.; Basso, A.; D'Arrigo, C.; Lunaccio, D.; Moni, L.; Riva, R.; Lambruschini, C. Multicomponent Synthesis of Polyphenols and Their in Vitro Evaluation as Potential β-Amyloid Aggregation Inhibitors. *Molecules* **2019**, *24*, 2636. [CrossRef] [PubMed]

3. Ren, X.; Bao, Y.; Zhu, Y.; Liu, S.; Peng, Z.; Zhang, Y.; Zhou, G. Isorhamnetin, Hispidulin, and Cirsimaritin Identified in Tamarix ramosissima Barks from Southern Xinjiang and Their Antioxidant and Antimicrobial Activities. *Molecules* **2019**, *24*, 390. [CrossRef] [PubMed]
4. Albouchi, F.; Avola, R.; Dico, G.M.L.; Calabrese, V.; Graziano, A.C.E.; Abderrabba, M.; Cardile, V. Melaleuca styphelioides Sm. Polyphenols Modulate Interferon Gamma/Histamine-Induced Inflammation in Human NCTC 2544 Keratinocytes. *Molecules* **2018**, *23*, 2526. [CrossRef] [PubMed]
5. Lizard, G.; Latruffe, N.; Vervandier-Fasseur, D. Aza- and Azo-Stilbenes: Bio-Isosteric Analogs of Resveratrol. *Molecules* **2020**, *25*, 605. [CrossRef] [PubMed]

© 2020 by the author. Licensee MDPI, Basel, Switzerland. This article is an open access article distributed under the terms and conditions of the Creative Commons Attribution (CC BY) license (http://creativecommons.org/licenses/by/4.0/).

Article

Synthesis of Bisphenol Neolignans Inspired by Honokiol as Antiproliferative Agents

Nunzio Cardullo [1,*], Vincenza Barresi [2], Vera Muccilli [1], Giorgia Spampinato [2], Morgana D'Amico [2], Daniele Filippo Condorelli [2] and Corrado Tringali [1,*]

[1] Department of Chemical Sciences, University of Catania, Viale A. Doria 6, 95125 Catania, Italy; v.muccilli@unict.it
[2] Department of Biomedical and Biotechnological Sciences, Section of Medical Biochemistry, University of Catania, Via Santa Sofia 97, 95123 Catania, Italy; vincenza.barresi@unict.it (V.B.); giorgiaspampinato@unict.it (G.S.); morganadamico01@gmail.com (M.D.); daniele.condorelli@unict.it (D.F.C.)
* Correspondence: ctringali@unict.it (C.T.); ncardullo@unict.it (N.C.); Tel.: +39-095-7385025 (C.T.)

Academic Editor: David Barker
Received: 15 January 2020; Accepted: 5 February 2020; Published: 7 February 2020

Abstract: Honokiol (2) is a natural bisphenol neolignan showing a variety of biological properties, including antitumor activity. Some studies pointed out 2 as a potential anticancer agent in view of its antiproliferative and pro-apoptotic activity towards tumor cells. As a further contribution to these studies, we report here the synthesis of a small library of bisphenol neolignans inspired by honokiol and the evaluation of their antiproliferative activity. The natural lead was hence subjected to simple chemical modifications to obtain the derivatives 3–9; further neolignans (12a-c, 13a-c, 14a-c, and 15a) were synthesized employing the Suzuki–Miyaura reaction, thus obtaining bisphenols with a substitution pattern different from honokiol. These compounds and the natural lead were subjected to antiproliferative assay towards HCT-116, HT-29, and PC3 tumor cell lines. Six of the neolignans show GI_{50} values lower than those of 2 towards all cell lines. Compounds 14a, 14c, and 15a are the most effective antiproliferative agents, with GI_{50} in the range of 3.6–19.1 μM, in some cases it is lower than those of the anticancer drug 5-fluorouracil. Flow cytometry experiments performed on these neolignans showed that the inhibition of proliferation is mainly due to an apoptotic process. These results indicate that the structural modification of honokiol may open the way to obtaining antitumor neolignans more potent than the natural lead.

Keywords: honokiol; bisphenol neolignans; polyphenols; Suzuki–Miyaura cross-coupling; antitumor activity; apoptosis

1. Introduction

The biaryl skeleton is relatively common among natural products and this structural feature is distinctive of bisphenol neolignans, a group of polyphenols belonging to the neolignan family [1]. These compounds are biosynthesized through oxidative coupling of phenoxy radicals generated by enzymes such as laccase, peroxidase, or a cytochrome P450 [2]. The most representative bisphenol neolignans are magnolol (1, Figure 1) and honokiol (2, Figure 1), originally isolated from roots and stem bark of *Magnolia officinalis* and *Magnolia obovata*. The extracts of *Magnolia* spp. (mainly *M. officinalis*) have been employed for centuries in traditional Chinese and Japanese medicine to treat many diseases, including anxiety, allergy, or gastrointestinal disorders [3,4]. These extracts have shown to possess promising biological activities, including anti-inflammatory, antioxidant, antiviral, anti-depressant, and anti-platelet activity [4,5].

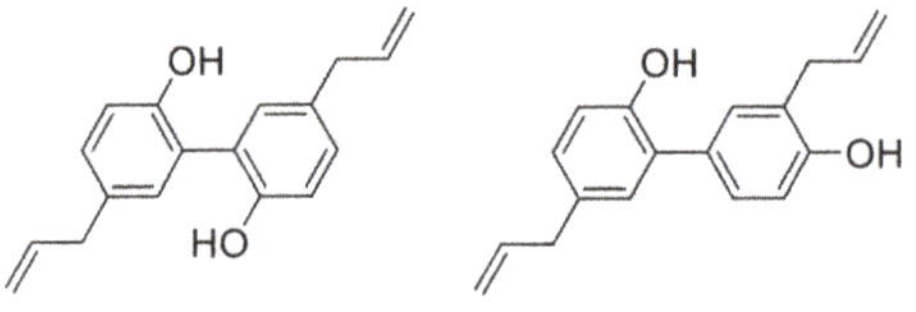

Figure 1. Chemical structures of magnolol (1) and honokiol (2).

Magnolol and honokiol are the main bioactive ingredients of these extracts [6] and have shown an array of biological properties, including antioxidant [6,7], anti-inflammatory [8], neuroprotective [9], and antitumor activity [10,11]. Specifically, 1 and 2 inhibit proliferation of tumor cells, inducing differentiation and apoptosis, and suppressing angiogenesis [12–15]. Furthermore, their unique pharmacophore structure, that is two phenolic rings linked through a C–C bond allows the interaction with a variety of biological targets [16].

The above-cited properties have prompted many researchers to synthesize magnolol and honokiol analogues and evaluate their biological properties to obtain new potential therapeutic agents. These efforts have afforded new bisphenol neolignans with optimized properties, among which antimicrobial [17,18], neuroprotective [19], anti-inflammatory [20], antitumor [18,21,22], and antiangiogenic activity [23]. According to some studies, the antitumor activity of honokiol, magnolol, and their analogues is related to the presence of free hydroxyl group and allylic chains on a bisphenolic moiety [18,23].

Although several synthetic methods have been employed to obtain biaryl compounds, the Pd-catalyzed Suzuki–Miyaura (S–M) cross-coupling reaction is one of the most efficient [24,25]. Moreover, with respect to other Pd-catalyzed reactions, S–M coupling has the advantage of requiring mild conditions and employing commercially available boronic acids that are environmentally safer than organometallic reagents [24]. On the other hand, oxidative coupling methods, based on the use of enzymes such as horseradish peroxidase [26], allow the synthesis of bisphenol neolignans in eco-friendly conditions but provide moderate or poor yield.

Thus, in continuation of our previous studies on the synthesis of natural-derived polyphenols with antitumor [27–31], antioxidant, hypoglycemic [32,33], antifungal [34], and anti-inflammatory activity [35,36], we oriented our recent works toward the synthesis of magnolol analogues, which were evaluated as potential antidiabetic [26], anticancer [37], and antioxidative [38] agents. As a further contribution, in the present work we report the synthesis of bisphenols neolignans inspired by honokiol (2). All the synthetic neolignans, in comparison with 2, have been evaluated for their potential antiproliferative activity towards three tumor cell lines, (HCT-116, HT-29 and PC3).

2. Results and Discussion

2.1. Synthesis

On the basis of the above-cited biological properties of the natural lead 2, we planned to synthesize a small library of honokiol-inspired bisphenol neolignans. A first group of honokiol analogues was obtained through simple modifications of 2, as depicted in Scheme 1. By acetylation and methylation, we obtained compounds 3 and 4, respectively. Peracetate derivatives usually undergo in vivo enzymatic hydrolysis and are frequently prepared to overcome the low metabolic stability and poor bioavailability of natural polyphenols [29,39], whereas methylated analogues of polyphenols have shown in many cases enhanced biological activity and high metabolic stability [39,40]. The neolignans 2, 3, and 4 were subjected to catalytic hydrogenation to give respectively 5, 6, and 7, as it is useful to establish the possible role of the terminal double bond. The spectroscopic data of compounds 3–5, and 7 were in agreement with those previously reported in the literature [41,42], whereas the new bisphenol neolignan

6 was subjected to spectroscopic characterization and the analysis of HRMS, ^{1}H and ^{13}C-NMR spectra confirmed the expected structure.

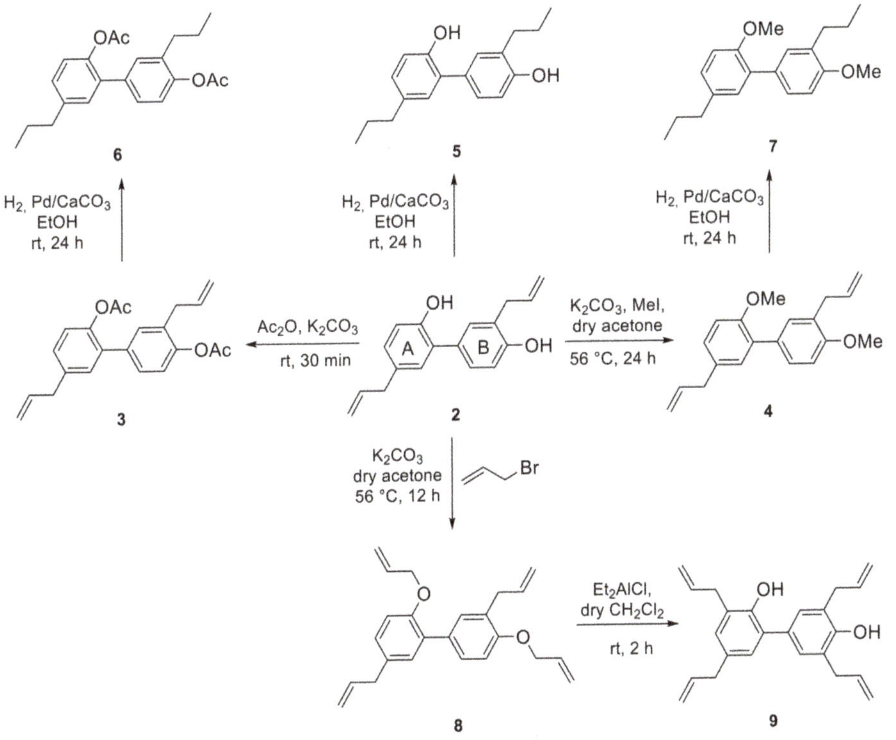

Scheme 1. Synthesis of honokiol derivatives 3–9.

According to the above cited report [18], the allylic chains on the bisphenolic core of honokiol are important structural requirements for antiproliferative activity; thus, we planned to investigate the effect of further allylic or *O*-allylic substituents. Namely, 2 was subjected to S$_N$2 reaction with allyl bromide to obtain the bis-*O*-allyl honokiol 8, whose structure was confirmed by analysis of its ^{1}H and ^{13}C-NMR data, in agreement with those previously reported [20]. As a further step, the Claisen rearrangement of 8 was planned to obtain the bis-*C*-allyl derivative 9. This reaction was carried out in mild conditions, namely at room temperature and in the presence of Et$_2$AlCl which catalyzes the [3,3]-sigmatropic rearrangement via an ether–aluminum complex, avoiding the use of high temperature. The analysis of ^{1}H and ^{13}C-NMR data, in agreement with those reported in literature, confirmed the structure of 9 [43].

Another set of bisphenol neolignans has been synthesized starting from commercial phenolics and employing the synthetic strategy reported in Schemes 2 and 3 based on the Suzuki–Miyaura cross-coupling reaction; accordingly, bisphenols 12a-c, 13a-c, 14a-c, and 15a were obtained.

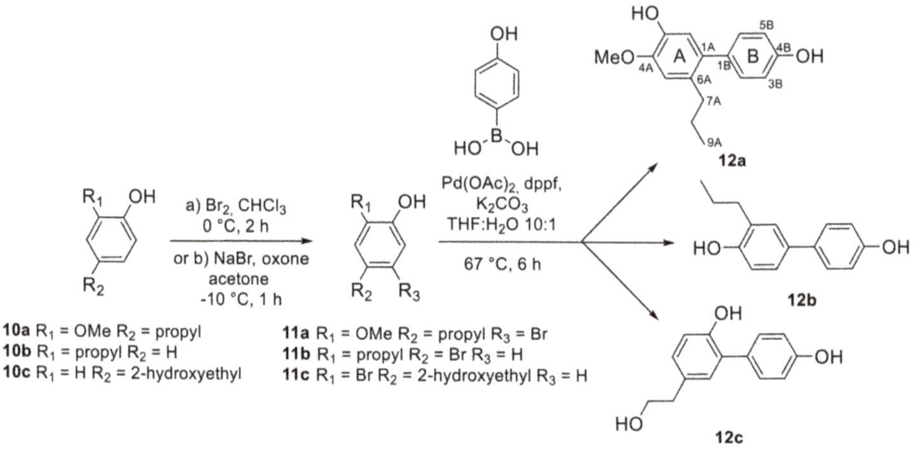

Scheme 2. Synthesis of bisphenol neolignans 12a-c. (**a**) These conditions were employed to obtain 11a and 11b; (**b**) these conditions were employed to obtain 11c.

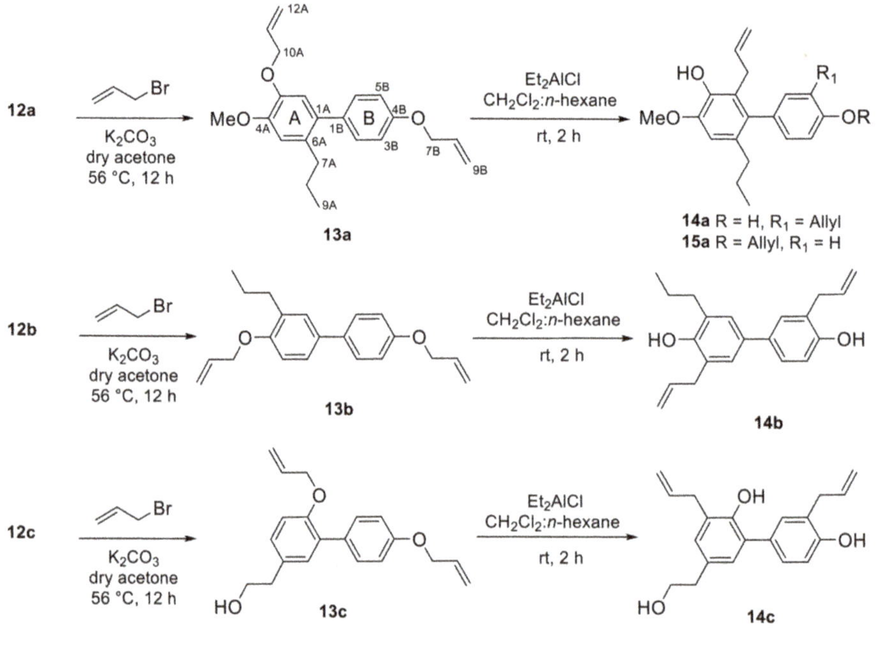

Scheme 3. Synthesis of bisphenol neolignans 13a-c, 14a-c, and 15a.

Schemes 2 and 3 summarize the final reaction conditions employed for each step, these were then optimized through a series of preliminary reactions. More specifically, the substrate 10a was used to optimize both the bromination and the S–M reactions. Preliminary experiments for the bromination were carried out employing three different brominating agents (namely, Br_2, N-bromosuccinimide, and NaBr/oxone) with or without a catalyst ($AlCl_3$ or I_2) and testing different solvents (CH_3CN, $CHCl_3$, and acetone). The reaction mixtures were analyzed by HPLC-UV on a C18 reversed-phase column in order to quantify the yield of the product 11a. These experiments are reported in detail in the experimental section and the results are summarized in Table 1.

Table 1. Reactions for bromination of 10a.

Entry	Brominating Agent	Catalyst	Solvent[1]	% Yield (11a)[2]
1	NBS	I_2	CH_3CN	10
2	NBS	I_2	$CHCl_3$	12
3	NBS	$AlCl_3$	CH_3CN	15
4	NBS	$AlCl_3$	$CHCl_3$	25
5	Br_2	$AlCl_3$	CH_3CN	18
6	Br_2	/	$CHCl_3$	47
7[3]	Br_2	/	$CHCl_3$	63
8[4]	NaBr/oxone	/	acetone/water	5

[1] If it is not indicated, the reactions were carried out at rt. [2] The yield was determined by HPLC-UV. [3] The reaction was carried out at 0 °C. [4] The reaction was performed at −10 °C.

When the substrate 10a was treated with Br_2 (entry 6) the expected monobromo derivative was obtained with higher yield (47%) respect to when the reaction occurred in other conditions. Furthermore, by working at 0 °C (entry 7) the yield grew up to 63%.

The same methodology was employed for the bromination of 10b, thus obtaining 11b with 67% yield; 10c afforded 11c with a lower yield (36%), hence, we applied the procedure previously described by Bovicelli et al. [44], namely by treating tyrosol (10c) with NaBr and oxone; in contrast with the low yield obtained for 11a (entry 8), 11c was recovered with 78% yield.

Also for the S–M coupling step a careful analysis of different reaction conditions was performed; the results for the coupling of 11a with 4-hydroxyphenyl boronic acid are summarized in Table 2, reporting the yields for the product 12a, subsequently established as the expected bisphenol neolignan (see below). The reaction was carried out varying solvent or solvent mixtures, the temperature or the bromide concentration (entries 3 and 4); $Pd(OAc)_2$ and the ligand 1,1'-bis(diphenylphosphino)ferrocene (dppf) were used to generate in situ the catalyst, and K_2CO_3 as base. The yield of neolignan 12a was determined by HPLC-UV analysis of the reaction mixtures. The results clearly indicated the best conditions for this step: a mixture THF/H_2O as solvent system at 70 °C, with 0.05 M concentration of bromide, affording 12a with 67% yield.

Table 2. Reactions for Suzuki–Miyaura cross coupling of 11a.

Entry	Solvent	Temperature	% Yield (12a)[1]
1	THF	25 °C	5
2	THF	70 °C	10
3	THF/H_2O[2]	70 °C	20
4	THF/H_2O[3]	70 °C	67
5	1,4-dioxane	70 °C	6
6	1,4-dioxane	180 °C	8

[1] The yield was determined by HPLC-UV. [2] The concentration of bromide 11a was 0.10 M. [3] The concentration of bromide 11a was 0.05 M.

On the basis of these encouraging results, the S–M coupling was carried out in a preparative scale and, after purification, 12a was submitted to a complete characterization by means of HRMS, ^1H and ^{13}C-NMR spectra analysis, including two-dimensional methods (COSY, HSQC, and HMBC). The HRMS spectrum confirmed the formation of a biphenyl structure. The NMR spectra showed the signals of a typical AA'XX' aromatic spin system assigned to ring B of 12a, namely two proton doublets at δ 7.14 (H-2B/H-6B) and 6.85 (H-3B/H-5B), with corresponding carbon signals at δ 130.7 (C-2B/C-6B) and 115.0 (C-3B/C-5B). Two sp^2 quaternary carbon signals were assigned at C-4B (δ 154.4) and C-1B (δ 134.4) on the basis of chemical shift and HMBC correlations with H-2B/H-6B and H-3B/H-5B. C-1B also showed a correlation with the singlet at δ 6.78, assigned to H-2A; this signal was HMBC correlated with the carbon at δ 134.6, assigned to C-1A on the basis of further HMBC correlations with H-5A and

H-2B/H-6B; C-1A was further correlated with the signal at δ 2.47, evidently due to the H_2-7A protons of the propyl chain. Overall, these and other HMBC data unambiguously established structure 12a.

The new bisphenol neolignan 12a was reacted with allyl bromide and afforded the bis-O-allyl derivative 13a, whose structure was confirmed by analysis of HRMS, ^{1}H and ^{13}C-NMR spectra. Namely, the mass spectrum proved that a double substitution occurred. This was of course confirmed by the NMR data clearly showing signals due to two allyl chains; these were distinguished on the basis of the HMBC correlations of C-3A and C-4B with the pertinent methylene signals in the ^{1}H-NMR spectrum.

As final step, the allyloxy neolignan 13a was used as substrate for a Claisen rearrangement (carried out as above reported for the preparation of 9) affording two main products. The HRMS and NMR data of the more polar product indicated that both allyl chains underwent the Claisen rearrangement. COSY and HMBC experiments corroborated this assumption, thus establishing structure 14a for this bisphenol neolignan. The less polar product showed ^{1}H and ^{13}C-NMR signals of ring B and those of one O-allyl chain substantially superimposable with those of 13a; the other signals indicated that the rearrangement occurred only for the ring A chain; thus, structure 15a was assigned to this product.

With the same protocol (Schemes 2 and 3) the new bisphenols 12b and c, 13b and c, and 14b and c were obtained and fully characterized.

2.2. Biochemical Assay

The honokiol derivatives 3–9 and the bioinspired bisphenol neolignans 12a-c, 13a-c, 14a-c, and 15a were evaluated as potential antiproliferative agents towards three tumor cell lines: HCT-116, HT-29 (both human colorectal adenocarcinoma) and PC3 (human prostate adenocarcinoma) employing the MTT colorimetric assay. The anticancer drug 5-fluorouracil (5-FU) was used as positive control, while honokiol (2) was included in the study for comparison. The results are reported in Table 3 as GI_{50} values (μM) and in Figure 2 for the sake of clarity. The majority of the tested compounds shows, at least on one cell line, a higher activity than that of the lead compound honokiol (GI_{50} = 18.2, 40.6 and 52.1 μM towards HCT-116, HT-29, and PC3 cells, respectively). In particular, six compounds (9, 12b, 14a-c, 15a) show GI_{50} values lower than those of 2 for all cell lines, with GI_{50} values in the range 3.6–47.9 μM. Also 5, the hydrogenated analogue of 2, shows GI_{50} values lower than those of the natural lead for both HT-29 and PC3 cell lines and a comparable value towards HCT-116. Finally, compound 3 is comparable to 2 towards both HCT-116 and PC3 cell lines. These results indicate that the structural modification of honokiol may open the way to obtaining antitumor neolignans more potent than the lead compound.

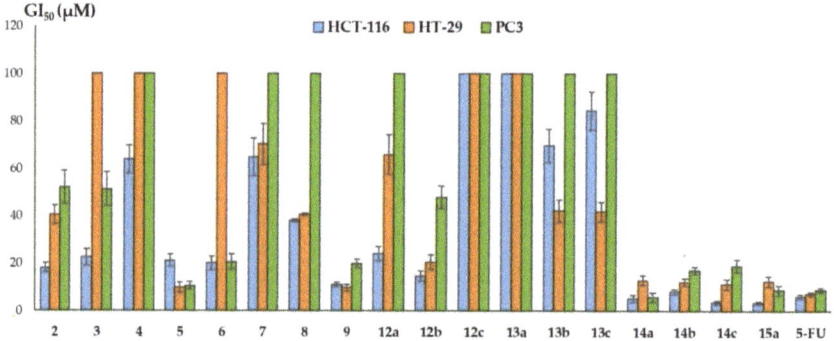

Figure 2. GI50 values (μM) of bisphenol neolignans 2–9, 12a-c, 13a-c, 14a-c, and 15a and of the reference compound 5-fluorouracil (5-FU) on HCT-116, HT-29, and PC3 cell lines after an incubation time of 72 h. The results shown are means ± SD of four experiments.

Table 3. Antiproliferative activity of bisphenol neolignans inspired by honokiol.

Compound	GI$_{50}$ (µM) ± SD [1]		
	HCT-116	HT-29	PC3
2	18.2 ± 2.1	40.6 ± 3.9	52.1 ± 7.1
3	22.5 ± 3.4	> 100	51.3 ± 7.2
4	63.9 ± 5.9	> 100	> 100
5	21.2 ± 2.6	9.9 ± 2.1	10.5 ± 1.7
6	20.1 ± 2.9	> 100	20.6 ± 3.2
7	64.7 ± 7.9	70.2 ± 8.7	> 100
8	38.1 ± 0.6	40.7 ± 0.5	> 100
9	11.1 ± 0.9	9.8 ± 1.4	19.8 ± 1.8
12a	24.1 ± 3.0	65.9 ± 8.3	> 100
12b	14.7 ± 2.1	20.5 ± 3.1	47.9 ± 4.7
12c	> 100	> 100	> 100
13a	> 100	> 100	> 100
13b	69.7 ± 6.9	42.3 ± 4.7	> 100
13c	84.5 ± 8.0	42.0 ± 4.1	> 100
14a	5.3 ± 1.5	13.0 ± 2.0	5.8 ± 1.9
14b	8.2 ± 1.1	12.3 ± 1.6	17.2 ± 1.5
14c	3.7 ± 0.7	11.3 ± 2.2	19.1 ± 2.6
15a	3.6 ± 0.6	12.7 ± 2.1	8.9 ± 2.0
5-FU	6.2 ± 0.8	7.3 ± 0.7	9.0 ± 0.9

[1] GI$_{50}$ value were calculated after 72 h of continuous exposure relative to untreated controls; values are the mean (±SD) of four experiments. HCT-116 and HT-29: human colorectal adenocarcinoma cells. PC3: human prostate cancer cells. 5-FU: 5-fluorouracil.

The bisphenol neolignans 14a, 14c, and 15a gave the most promising results, in particular 14a and 15a showed antiproliferative activity higher or comparable with that of the anticancer drug 5-FU against both HCT-116 and PC3.

Although the data reported in Table 3 do not allow conclusive assessments about the structural determinants required for an optimized antiproliferative activity of honokiol-inspired neolignans analogues, some considerations about structure-activity relationships can be made and are reported below.

The neolignans 9, 12b, 14a-c, and 15a, with higher antiproliferative activity than 2, possess one allyl or propyl chain in ortho position to a free phenolic group. Among these, 14a, 14b and 14c present the same structural motif of honokiol on ring B. In particular, the presence of free phenolic groups seems to be a pivotal requirement: in fact, compound 4, the dimethyl ether of honokiol, is practically inactive, and the majority of poorly or not active neolignans have no free phenolic groups, with the exception of 12a and 12c, lacking of the allyl chains present in honokiol. Diacetates 3 and 6 show an activity slightly lower than that of 2 towards HCT-116 and PC3 cells, suggesting that these compounds may act as prodrug and may release 2 or the hydrogenated honokiol in presence of intracellular esterases. The above cited structural features are not present in compounds with very low antiproliferative activity. The presence of one or more allyloxy groups does not seem to be, by itself, an essential structural motif, being present in active compounds such as 15a, but also in poor or not active neolignans, such as 8, 13a-c. Finally, it is worthy of note that the hydrogenation of honokiol to give 5 causes an enhancement of activity toward HT-29 and PC3 cells, thus suggesting that these terminal double bonds are not essential for the activity.

On the basis of the above data, we selected three of the most potent neolignans, namely 14a, 14c, and 15a for a flow cytometric analysis on HCT-116 and PC3 cells. This analysis showed that the inhibition of proliferation is mainly due to an apoptotic process, with high values of apoptotic cells in almost all assays (Table 4). Cells treated with 15a showed the highest values of apoptotic cells in both lines: 53.3% of HCT-116 and 38.4% of PC3 were detected in early and late apoptosis status (Figure 3, Table 4). On the contrary, necrotic cells detectable by Propidium Iodide (PI) staining alone were not revealed at significant levels for all tested compounds in both cell lines (Table 4 and Figure 3 for 15a). The results obtained by antiproliferative assay and flow cytometry suggest that selected bioinspired bisphenol neolignans could be further examined in depth in order to define the underlying antitumor mechanism.

Table 4. Apoptotic death of 14a, 14c, and 15a in HCT-116 and PC3 Cells[1].

Entry	HTC-116 Cell Distribution (%)					PC3 Cell Distribution (%)				
	Ctrl	14a	14c	15a	5-FU	Ctrl	14a	14c	15a	5-FU
Live	82.1	74.8	70.9	46.3	17.0	92.7	79.9	86.5	61.3	24.0
Early apoptosis (Annexin+)	1.5	9.5	6.0	17.4	55.0	0.9	12.4	5.3	26.0	50.0
Necrotic cells (PI+)	1.6	4.8	4.4	0.4	0.2	1.7	0.1	0.2	0.4	0.3
Late apoptosis (Annexin+/PI+)	14.9	10.9	18.7	35.9	27.8	4.7	7.7	8.0	12.4	25.7

[1] Determined by Alexa Fluor 488; annexin V/propidium iodide (annexin/PI) staining after treatment with 14a, 14c, 15a, and 5-FU (10 µM) for 72 h. The analysis was performed on 10,000 events for each condition and expressed in percentage of total number of events.

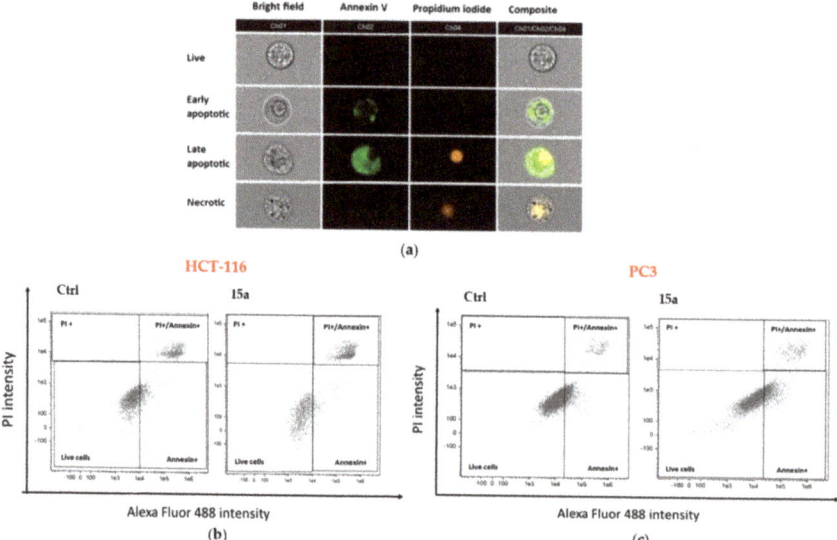

Figure 3. Flow cytometry: (a) Typical images of cells analyzed by flow cytometry (Amnis FlowSigh). Each cell (event) is visible in a bright field and stained by annexin-V positive, propidium iodide positive, and double positive cells; Flow cytometric dot plot of specific cell populations in HCT-116 (b) and PC3 (c) cell lines in the presence of 15a: live (double annexin/PI negative), necrosis (annexin negative and PI positive), early apoptosis (annexin positive), and late apoptosis (double annexin/PI positive).

3. Materials and Methods

3.1. General Information

All chemicals were of reagent grade, and were used without further purification. Honokiol, 4-hydroxyphenylboronic acid, 1,1′-bis(diphenylphosphino)ferrocene (dppf), Pd(OAc)$_2$ were purchased from TCI Europe (Milan, Italy), eugenol, 2-allyl phenol, tyrosol were purchased from Sigma Aldrich (Milan, Italy).

Preparative liquid chromatography was performed on silica gel (63-200 µm, Merck, Darmstadt, Germany), or Sephadex-LH20 (Sigma-Aldrich, Milan, Italy) using different mixtures of solvents, as reported for each compound. TLC was carried out using pre-coated silica gel F254 plates

(Macherey-Nagel, Düren, Germany); visualization of reaction components was achieved under UV light at wavelengths of 254 and 366 nm, and by staining with a solution of cerium sulfate followed by heating.

HPLC-UV instrument (Agilent, Milan, Italy), equipped with an auto-sampler (G1313A), a pump (G1354A) and a diode array detector (DAD; G1315D), was employed for quantitative analysis with an analytical reversed-phase column (Luna C18, 5 μm; 4.6 × 250 mm; Phenomenex, Castel Maggiore, BO, Italy) and eluting at 1 mL/min with the following gradient of CH_3CN–HCOOH (99:1 v/v; A) in H_2O–HCOOH (99:1 v/v; B): t_0 min A = 50%, t_{15} min A = 100%, t_{20} min A = 50%.

NMR spectra were run on a Varian Unity Inova spectrometer (Italy, Milan) operating at 499.86 (^{1}H) and 125.70 MHz (^{13}C), and equipped with a gradient-enhanced, reverse-detection probe. Chemical shifts (δ) are indirectly referred to TMS using residual solvent signals. All NMR experiments, including 2D spectra, i.e., g-COSY, g-HSQCAD, and g-HMBCAD, were performed using software supplied by the manufacturer, and acquired at constant temperature (300 K). g-HMBCAD experiments were optimized for a long-range ^{13}C-^{1}H coupling constant of 8.0 Hz. High-resolution mass spectra were acquired with an Orbitrap Fusion Tribrid®(Q-OT-qIT) mass spectrometer (Thermo Fisher Scientific, Bremen, Germany) equipped with an ESI ion source operating in positive or negative mode. Samples were directly infused and converted to the gas phase using the following parameters: source voltage, 2.6 kV; sheath gas flow rate, 25 au; and auxiliary gas, 8 au. The ions were introduced into the mass spectrometer through a heated ion transfer tube (300 °C). Survey scan was performed from m/z 150 to 1000 at 500k resolution (@ 200 m/z) using the following parameters: RF lens, 60%; auto gain control target, 20,000.

3.2. Synthesis of Compound 3

Honokiol (214.2 mg, 0.76 mmol) was mixed with acetic anhydride (10 mL) and with K_2CO_3 (445.3 mg, 3.2 mmol) at rt for 30 min. The mixture was diluted with cold water (10 mL) and partitioned with EtOAC (3 × 15 mL), the combined organic phase was dried over anhydrous Na_2SO_4, filtered and taken to dryness. The recovered organic layer was purified by column chromatography on silica gel (n-hexane → n-hexane:acetone 60:40) to give acetyl honokiol (3) with 56% yield (150.4 mg). Spectroscopic data were in agreement with those previously reported [41].

3.3. Synthesis of Compound 4

Honokiol (200 mg, 0.75 mmol) was solubilized in dry acetone (15 mL) and the solution was mixed with K_2CO_3 (630.3 mg, 4.5 mmol) for 10 min. Then, MeI was added (0.19 mL, 3.2 mmol) and the mixture was refluxed for 24 h. The mixture was filtered out and the permethylated compound 4 was recovered after column chromatography on silica gel (cyclohexane → cyclohexane:acetone 80:20) with 84% yield (185.6 mg). Spectroscopic data were in agreement with those previously reported [42].

3.4. Synthesis of Compounds 5–7, 10a and b

The hydrogenation of 2–4, eugenol, and 2-allyl phenol (0.5 mmol) was carried out employing absolute EtOH (7 mL) and Pd/CaCO$_3$ (10% w/w; 17.4 mg) as catalyst. The reaction flask was filled with H_2 (1 atm) and stirred at rt for 24 h. The catalyst was removed by filtration on Celite 545. In these conditions the expected products have been obtained quantitatively without further purification. The spectroscopic data of 3′,5-dipropyl-(1,1′-biphenyl)-2,4′-diol (5) [42] 2,4′-dimethoxy-3′,5-dipropyl-1,1′-biphenyl (7) [42] 2-methoxy-4-propylphenol (10a) [45] and 2-propylphenol (10b) [46] were in agreement with literature data.

3′,5-Dipropyl-(1,1′-biphenyl)-2,4′-diyl diacetate (6). Yield: 96% (169.7 mg). R_f 0.47 (n-hexane:acetone 70:30). ^{1}H-NMR (CDCl$_3$): δ 7.29 (d, J = 2.0 Hz, 1 H, H-2B), 7.25 (dd, J = 8.5, 2.0 Hz, 1 H, H-6B), 7.20 (d, J = 1.9 Hz, 1 H, H-6A), 7.17 (dd, J = 8.2, 1.9 Hz, 1 H, H-4A), 7.06 (d, J = 8.5 Hz, 1 H, H-5B), 7.03 (d, J = 8.2 Hz, 1 H, H-3A), 2.62 (bdd, J = 7.5 Hz, 2 H, CH$_2$-7A), 2.54 (bdd, J = 7.6 Hz, 2 H, CH$_2$-7B), 2.34 (s, 3 H, OCOCH$_3$), 2.09 (s, 3 H, OCOCH$_3$), 1.66 (m, 4 H, CH$_2$-8A and CH$_2$-8B), 0.96 (m, 6 H, CH$_3$-9A and CH$_3$-9B). ^{13}C-NMR (CDCl$_3$): δ 169.8 (C, OCOCH$_3$), 169.6 (C, OCOCH$_3$), 148.5 (C, C-4B), 145.8 (C, C-2A), 140.9 (C, C-5A), 135.7 (C, C-3B), 134.1 (C, C-1B), 133.9 (C, C-1A), 131.0 (CH, C-4A), 130.9 (CH, C-2B), 128.6 (CH, C-6A), 127.5 (CH, C-6B), 122.6 (CH, C-3A), 122.2 (CH, C-5B), 37.6 (CH$_2$, C-7A), 32.4 (CH$_2$, C-7B), 24.6 (CH$_2$, C-8A), 23.3 (CH$_2$, C-8B), 21.1 (CH$_3$, OCOCH$_3$), 21.0 (CH$_3$, OCOCH$_3$), 14.2 (CH$_3$, C-9B), 14.0 (CH$_3$, C-9A). HRMS (ESI+) m/z 377.1739 [M+Na]$^+$ (calcd for C$_{22}$H$_{26}$O$_4$Na, 377.1728).

3.5. Synthesis of Bromophenols 11a-c

Preliminary experiments for bromination have been performed and the details of these experiments have been reported in Supplementary Materials.

According to the experiment 7 reported in Supplementary Materials (entry 7 of Table 1), a solution of each compound (10a-b; 5 mmol) in CHCl$_3$ (17 mL) was kept in ice bath and a solution containing Br$_2$ (300 μL; 1.2 mmol in 10 mL CHCl$_3$), was added dropwise. The reaction was monitored by TLC and the Br$_2$ was quenched by the addition of a saturated Na$_2$S$_2$O$_3$ solution (15 mL). The mixture was partitioned with CH$_2$Cl$_2$ (3 x 15 mL) and the organic layer was dried and taken to dryness.

5-Bromo-2-methoxy-4-propylphenol (11a). The expected compound was recovered by column chromatography on silica gel (cyclohexane:EtOAc 98:2 → cyclohexane:EtOAc 96:4) with 63% yield (765.2 mg). R$_f$ 0.48 (cyclohexane:EtOAc 75:25).

4-Bromo-2-propylphenol (11b). The organic layer was purified on silica gel column chromatography (cyclohexane → cyclohexane:EtOAc 95:5) obtaining 11b with 67% yield (720.1 mg). R$_f$ 0.31 (n-hexane:acetone 70:30).

2-Bromo-4-(2-hydroxyethyl)phenol (11c). Compound 11c was obtained as previously described [44]. Briefly, a solution of tyrosol (10c; 340.3 mg; 2.5 mmol) in acetone (9.2 mL) was stirred with NaBr (514.3 mg; 5 mmol) at −10 °C and a 0.33 M oxone solution (2 gr in 10 mL of H$_2$O) was dropwise added. The mixture was stirred for 1 h at −10 °C and then it was partitioned with EtOAc (3 x 10 mL). The combined organic layer was dried over anhydrous Na$_2$SO$_4$, filtered and taken to dryness. The column chromatography on silica gel (cyclohexane → cyclohexane: acetone 65:35) afforded 11c with 78% yield (423.5 mg).

NMR data of the isolated compounds are in agreement with those previously reported for 11a [47], 11b [48], and 11c [44].

3.6. Suzuki–Miyaura Cross-Coupling Reaction: Synthesis of Bisphenols 12a-c

The experimental conditions for the preliminary experiments performed on 11a for S–M reaction have been reported in Supplementary Materials.

According to entry 4 of these experiments (entry 4 in Table 2), each aryl bromide 11a-c (1.0 mmol) was solubilized in THF (17 mL) and mixed with 4-hydroxyphenylboronic acid (207.2 mg, 1.5 mmol), dppf (165.7 mg, 0.3 mmol), Pd(OAc)$_2$ (22.5 mg, 0.1 mmol). Then, a 3 M K$_2$CO$_3$ solution (1.7 mL, 5.0 mmol) was added and the mixture was stirred at 70 °C for 6 h. The mixture was diluted with water (20 mL) and partitioned with EtOAc (3 x 25 mL). The combined organic layer was washed, dried over anhydrous Na$_2$SO$_4$, filtered and taken to dryness. The expected bisphenol was recovered after column chromatography.

4-Methoxy-6-propyl-(1,1'-biphenyl)-3,4'-diol (12a). The silica gel column chromatography (petroleum ether → petroleum ether:acetone 92:8) furnished the bisphenol neolignan 12a with 65% yield (167.7 mg). R$_f$ 0.39 (cyclohexane:acetone 70:30). ^{1}H-NMR (CDCl$_3$): δ 7.14 (d, J = 8.5 Hz, 2 H, H-2B/H-6B), 6.85 (d, J = 8.5 Hz, 2 H, H-3B/H-5B), 6.78 (s, 1 H, H-2A), 6.76 (s, 1 H, H-5A), 5.52 (bs, 1 H, 3A-OH), 5.07 (bs, 1 H, 4B-OH), 3.92 (s, 3 H, OCH$_3$), 2.47 (m, 2 H, CH$_2$-7A), 1.47 (m, 2 H, CH$_2$-8A), 0.82 (t, J = 7.4 Hz, 3 H, CH$_3$-9A). ^{13}C-NMR (CDCl$_3$): δ 154.4 (C, C-4B), 145.7 (C, COCH$_3$), 143.1 (C, C-3A), 134.6 (C, C-1A), 134.4 (C, C-1B), 132.2 (CH, C-6A), 130.7 (CH, C-2B/C-6B), 116.4 (CH, C-2A), 115.0 (CH, C-3B/C-5B), 111.7 (CH, C-5A), 56.1 (CH$_3$, OCH$_3$), 35.0 (CH$_2$, C-7A), 25.0 (CH$_2$, C-8A), 14.1 (CH$_3$, C-9A). HRMS (ESI-) m/z 257.1169 [M-H]$^-$ (calcd for C$_{16}$H$_{17}$O$_3$, 257.1178).

3-Propyl-(1,1'-biphenyl)-4,4'-diol (12b). The silica gel column chromatography (*n*-hexane:acetone 95:5 → 88:12) afforded the bisphenol neolignan 12b with 70% yield (158.2 mg). R_f 0.35 (cyclohexane:acetone 70:30). ^{1}H-NMR ((CD$_3$)$_2$CO): δ 7.40 (d, *J* = 8.4 Hz, 2 H, H-2B/H-6B), 7.31 (d, *J* = 1.9 Hz, 1 H, H-2A), 7.22 (dd, *J* = 8.2, 1.9 Hz, 1 H, H-6A), 6.88 (d, *J* = 8.4 Hz, 2 H, H-3B/H-5B), 6.87 (d, *J* = 8.2 Hz, 1 H, H-5A), 2.64 (m, 2 H, CH$_2$-7A), 1.67 (m, 2 H, CH$_2$-8A), 0.98 (t, *J* = 7.4 Hz, 3 H, CH$_3$-9A). ^{13}C-NMR ((CD$_3$)$_2$CO): δ 157.1 (C, C-4B), 145.9 (C, C-4A), 133.6 (C, C-1B), 133.2 (C, C-1A), 129.7 (C, C-3A), 129.0 (CH, C-6A), 128.2 (CH, C-3B/C-5B), 125.5 (C, C-2A), 116.3 (CH, C-2B/C-6B), 116.0 (CH, C-5A), 33.1 (CH$_2$, C-7A), 23.8 (CH$_2$, C-8A), 14.3 (CH$_3$, C-9A). HRMS (ESI-) *m/z* 227.1065 [M-H]$^-$ (calcd for C$_{15}$H$_{15}$O$_2$, 227.1072).

5-(2-Hydroxyethyl)-(1,1'-biphenyl)-2,4'-diol (12c). The neolignan 12c was recovered after column chromatography (cyclohexane.acetone 98:2 → 70:30) with 68% yield (155.7 mg) R_f 0.2 (cyclohexane:acetone 60:40). ^{1}H-NMR ((CD$_3$)$_2$CO): δ 8.30 (bs C-4B-O*H*), 7.91 (bs, C-2A-O*H*), 7.43 (d, *J* = 8.6 Hz, 2 H, H-2B/H-6B), 7.12 (d, *J* = 2.2 Hz, 1 H, H-6A), 6.99 (dd, *J* = 8.2, 2.2 Hz, 1 H, H-4A), 6.87 (d, *J* = 8.6 Hz, 2 H, H-3B/H-5B), 6.85 (d, *J* = 8.2 Hz, 1 H, H-3A), 3.75 (t, *J* = 7.1 Hz, 2 H, CH$_2$-8A), 3.53 (bs, 1 H, O*H*-8A), 2.76 (t, *J* = 7.1 Hz, 2 H, CH$_2$-7A). ^{13}C-NMR ((CD$_3$)$_2$CO): δ 157.2 (C, C-4B), 153.1 (C, C-2A), 131.7 (CH, C-6A), 131.6 (C, C-5A), 131.2 (CH, C-2B/C-6B), 131.0 (C, C-1B), 129.1 (CH, C-4A), 129.0 (C, C-1A), 116.7 (CH, C-3A), 115.7 (CH, C-3B/C-5B), 64.2 (CH$_2$, C-8A), 39.5 (CH$_2$, C-7A). HRMS (ESI-) S *m/z* 229.0874 [M-H]$^-$ (calcd for C$_{14}$H$_{13}$O$_3$, 229.0865).

3.7. Synthesis of O-Allyloxy Neolignans 8 and 13a-c

A solution of the proper bisphenol neolignan (2, 12a-c; 0.5 mmol) in dry acetone (5 mL) was mixed with K$_2$CO$_3$ (275.7 mg; 2.0 mmol) for 10 min, then, allyl bromide (130 µL; 1.5 mmol) was added and the mixture was refluxed overnight. The mixture was filtered and where necessary it was chromatographed.

3',5-Diallyl-2,4'-bis(allyloxy)-1,1'-biphenyl (8). The expected compound was recovered with 97% yield (184.9 mg) after filtration of the mixture. R_f 0.79 (cyclohexane:acetone 70:30). Spectroscopic data were in agreement with those reported in the literature [20].

4',5-Bis(allyloxy)-4-methoxy-2-propyl-1,1'-biphenyl (13a). The mixture was purified on Sephadex-LH20 column chromatography eluting with CH$_2$Cl$_2$, to give 13a with 95% yield (160.5 mg). R_f 0.69 (cyclohexane:acetone 70:30). ^{1}H-NMR (CDCl$_3$): δ 7.19 (d, *J* = 8.6 Hz, 2 H, H-2B/H-6B), 6.94 (d, *J* = 8.6 Hz, 2 H, H-3B/H-5B), 6.78 (s, 1 H, H-5A), 6.73 (s, 1 H, H-2A), 6.11 (m, 1 H, H-11A), 6.07 (m, 1 H, H-8B), 5.45 (dd, *J* = 15.0, 5.0 Hz, 1 H, CH$_a$-12A), 5.39 (dd, *J* = 15.0, 5.0 Hz, 1 H, CH$_a$-9B), 5.31 (dd, *J* = 10.0, 5.0 Hz, 1 H, CH$_b$-12A), 5.26 (dd, *J* = 10.0, 5.0 Hz, 1 H, CH$_b$-9B), 4.59 (d, *J* = 5.3 Hz, 2 H, CH$_2$-7B), 4.58 (d, *J* = 5.4 Hz, 2 H, CH$_2$-10A), 3.91 (s, 3 H, OCH$_3$), 2.48 (m, 2 H, CH$_2$-7A), 1.49 (m, 2 H, CH$_2$-8A), 0.32 (t, *J* = 7.3 Hz, 3 H, CH$_3$-9A). ^{13}C-NMR (CDCl$_3$): δ 157.7 (C, C-4B), 148.5 (C, COCH$_3$), 145.7 (C, C-3A), 134.6 (C, C-1B), 133.8 (C, C-1A), 133.7 (CH, C-11A), 133.5 (CH, C-8B), 133.1 (C, C-6A), 130.5 (CH, C-2B/C-6B), 117.9 (CH$_2$, C-9B), 117.8 (CH$_2$, C-12A), 115.8 (CH, C-2A), 114.4 (CH, C-3B/C-5B), 112.9 (CH, C-5A), 70.1 (CH$_2$, C-10A), 69.0 (CH$_2$, C-7B), 56.2 (CH$_3$, OCH$_3$), 35.0 (CH$_2$, C-7A), 24.9 (CH$_2$, C-8A), 14.2 (CH$_3$, C-9A). HRMS (ESI+) *m/z* 361.1791 [M+Na]$^+$ (calcd for C$_{22}$H$_{26}$O$_3$Na, 361.1780).

4,4'-Bis(allyloxy)-3-propyl-1,1'-biphenyl (13b). The expected compound was recovered after filtration without further purification with 96% yield (147.2 mg). R_f 0.79 (cyclohexane:acetone 75:25). ^{1}H-NMR (CDCl$_3$): δ 7.47 (d, *J* = 8.7 Hz, 2 H, H-2B/H-6B), 7.33 (d, *J* = 1.9 Hz, 1 H, H-2A), 7.31 (dd, *J* = 8.2, 1.9 Hz, 1 H, H-6A), 6.97 (d, *J* = 8.7 Hz, 2 H, H-3B/H-5B), 6.87 (d, *J* = 8.2 Hz, 1 H, H-5A), 6.09 (m, 2 H, H-11A and H-8B), 5.45 (m, 2H, CH$_2$-12A), 5.30 (m, 2 H, CH$_2$-9B), 4.58 (d, *J* = 5.1 Hz, 4 H, CH$_2$-10A and CH$_2$-7B), 2.68 (t, *J* = 7.5 Hz, 2 H, CH$_2$-7A), 1.68 (sext, *J* = 7.5 Hz, 2 H, CH$_2$-8A), 0.98 (t, *J* = 7.5, 3 H, CH$_3$-9A). ^{13}C-NMR (CDCl$_3$): δ 157.8 (C, C-4B), 155.8 (C, C-4A), 134.1 (C, C-1B), 133.3 (C, C-1A), 133.8 (CH, C-11A), 133.5 (CH, C-8B), 131.8 (C, C-3A), 128.6 (CH, C-2A), 127.9 (CH, C-2B/C-6B), 125.0 (CH, C-6A), 117.8 (CH$_2$, C-9B), 116.9 (CH$_2$, C-12A), 115.1 (CH, C-3B/C-5B), 112.0 (CH, C-5A), 69.1 (CH$_2$, C-10A), 68.1 (CH$_2$, C-7B), 32.7 (CH$_2$, C-7A), 23.3 (CH$_2$, C-8A), 14.3 (CH$_3$, C-9A). HRMS (ESI+) *m/z* 331.1683 [M+Na]$^+$ (calcd for per C$_{21}$H$_{24}$O$_2$Na, 331.1674).

5-(2-Hydroxyethyl)-2,4′-bis(allyloxy)-1,1′-biphenyl (13c). The Sephadex-LH20 column chromatography (CH$_2$Cl$_2$) afforded the expected *O*-allyl derivative 13c with 75% yield (116.8 mg). R$_f$ 0.49 (cyclohexane.acetone 65:35). ^{1}H-NMR ((CD$_3$)$_2$CO): δ 7.49 (d, *J* = 8.8 Hz, 2 H, H-2B/H-6B), 7.18 (d, *J* = 1.9 Hz, 1 H, H-6A), 7.13 (dd, *J* = 8.3, 1.9 Hz, 1 H, H-4A), 6.97 (d, *J* = 8.7 Hz, 2 H, H-3B/H-5B), 6.96 (d, *J* = 8.3 Hz, 1 H, H-3A), 6.10 (m, 1 H, H-11A), 6.01 (m, 1 H, H-8B), 5.44 (dd, *J* = 17.3, 1.5 Hz, 1 H, CH$_a$-12A), 5.34 (dd, *J* = 17.3, 1.7 Hz, 1 H, CH$_a$-9B), 5.25 (dd, *J* = 10.6, 1.5 Hz, 1 H, CH$_b$-12A), 5.16 (dd, *J* = 10.6, 1.7 Hz, 1 H, CH$_b$-9B), 4.61(d, *J* = 5.2 Hz, 2 H, CH$_2$-10A), 4.53 (d, *J* = 3.2 Hz, 2 H, CH$_2$-7B), 3.75 (t, *J* = 6.6 Hz, 2 H, CH$_2$-8A), 2.79 (t, *J* = 6.6 Hz, 2 H, CH$_2$-7A). ^{13}C-NMR ((CD$_3$)$_2$CO): δ 158.6 (C, C-4B), 154.8 (C, C-2A), 134.88 (CH, C-11A), 134.86 (CH, C-8B), 133.2 (C, C-5A), 132.2 (C, C-1B), 131.9 (CH, C-6A), 131.4 (CH, C-2B/C-6B), 131.1 (C, C-1A), 129.4 (CH, C-4A), 117.3 (CH$_2$, C-9B), 116.7 (CH, C-3A), 114.9 (CH, C-3B/C-5B), 113.9 (CH$_2$, C-12A), 69.8 (CH$_2$, C-10A), 69.3 (CH$_2$, C-7B), 64.1 (CH$_2$, C-8A), 39.5 (CH$_2$, C-7A). HRMS (ESI-) *m/z* 309.1501 [M-H]$^-$ (calcd for C$_{20}$H$_{21}$O$_3$, 309.1491).

3.8. General Procedure for Claisen Rearrangement: Synthesis of Bisphenols 14a-c and 15a

A 1 M Et$_2$AlCl solution (in *n*-hexane; 1.5 mL) was added dropwise to a solution of *O*-allyl derivatives 8, 13a-c (0.35 mmol) in dry CH$_2$Cl$_2$ (2 mL). The mixture was stirred at rt for 2 h and then the reaction was quenched adding 2 N HCl solution (5 mL). The mixture was partitioned with CH$_2$Cl$_2$ (2 × 5 mL); the combined organic phase was washed with water, dried oved anhydrous Na$_2$SO$_4$, filtered and taken to dryness. Column chromatography on Sephadex-LH20 (CH$_2$Cl$_2$) afforded the pure products.

3,3′,5,5′-Tetraallyl-(1,1′-biphenyl)-2,4′-diol (9). Yield: 20% (24.7 mg). R$_f$ 0.55 (cyclohexane.acetone 70:30). Spectroscopic data were in agreement with those previously reported [43].

2,3′-Diallyl-4-methoxy-6-propyl-(1,1′-biphenyl)-3,4′-diol (14a). Yield: 18% (21.5 mg). R$_f$ 0.44 (cyclohexane:acetone 70:30). ^{1}H-NMR (CDCl$_3$): δ 6.88 (m, 1 H, H-6B), 6.87 (bs, 1 H, H-2B), 6.81 (d, *J* = 7.9 Hz, 1 H, H-5B), 6.66 (s, 1 H, H-5A), 6.03 (m, 1 H, H-8B), 5.82 (m, 1 H, H-11A), 5.59 (bs, 1H, 3A-OH), 5.13 (m, 2 H, CH$_2$-9B), 4.96 (bs, 1 H, 4B-OH), 4.87 (bd, *J* = 10.1 Hz, 1 H, CH$_a$-12A), 4.75 (bd, *J* = 17.1 Hz, 1 H, CH$_b$-12A), 3.91 (s, 3 H, OCH$_3$), 3.41 (d, *J* = 6.2 Hz, 2 H, CH$_2$-7B), 3.12 (d, *J* = 6.1 Hz, 2 H, CH$_2$-10A), 2.24 (bdd, *J* = 7.5 Hz, 2 H, CH$_2$-7A), 1.40 (sext, *J* = 7.5 Hz, 2 H, CH$_2$-8A), 0.77 (t, *J* = 7.5 Hz, 3 H, CH$_3$-9A). ^{13}C-NMR (CDCl$_3$): δ 152.7 (C, C-4B), 145.2 (C, COCH$_3$), 141.4 (C, C-3A), 136.8 (CH, C-11A), 136.4 (CH, C-8A), 134.5 (C, C-1A), 132.6 (C, C-6A), 132.5 (C, C-1B), 132.3 (CH, C-2B), 129,4 (CH, C-6B), 124.5 (C, C-2A), 124.4 (C,C-3B), 116.3 (CH, C-5B), 115.2 (CH$_2$, C-9B), 114.4 (CH$_2$, C-12A), 109.1 (CH, C-5A), 55.9 (CH$_3$, OCH$_3$), 35.8 (CH$_2$, C-7A), 34.9 (CH$_2$, C-7B), 32.1 (CH$_2$, C-10A), 24.9 (CH$_2$, C-8A), 14.0 (CH$_3$, C-9A). HRMS (ESI-) *m/z* 337.2123 [M-H]$^-$ (calcd for C$_{22}$H$_{25}$O$_3$, 337.2117).

3,3′-Diallyl-5-propyl-(1,1′-biphenyl)-4,4′-diol (14b). Yield: 40% (42.8 mg) R$_f$ 0.48 (cyclohexane:acetone 70:30). ^{1}H-NMR (CDCl$_3$): δ 7.31 (dd, *J* = 8.2, 2.1 Hz, 1 H, H-6B), 7.28 (d, *J* = 2.1 Hz, 1 H, H-2B), 7.19 (d, *J* = 1.5 Hz, 1 H, H-2A), 7.13 (d, *J* = 1.5 Hz, 1 H, H-6A), 6.85 (d, *J* = 8.2 Hz, 1 H, H-5B), 6.06 (m, 2 H, H-11A and H-8B), 5.21 (m, 4 H, CH$_2$-12A and CH$_2$-9B), 4.98 (bs, 1H, 4A-OH) 4.96 (bs, 1 H, 4B-OH), 3.47 (d, *J* = 6.3 Hz, 4 H, CH$_2$-10A and CH$_2$-7B), 2.73 (bt, *J* = 7.7 Hz, 2 H, CH$_2$-7A), 1.68 (sext, *J* = 7.5 Hz, 2 H, CH$_2$-8A), 1.0 (t, *J* = 7.5 Hz, 3 H, CH$_3$-9A). ^{13}C-NMR (CDCl$_3$): δ 153.3 (C, C-4B), 151.7 (C, C-4A), 136.6 (CH, C-11A and C-6A), 136.5 (CH, C-8B), 134.3 (C, C-1B), 133.4 (C, C-1A), 129,4 (C, C-3A), 127.2 (CH, C-2A), 126.4 (CH, C-6B), 125.5 (C, C-3B), 125.0 (C, C-5A), 117.0 (CH$_2$, C-12A), 116.8 (CH$_2$, C-9B), 116.3 (CH, C-5B), 36.1 (CH$_2$, C-10A), 35.6 (CH$_2$, C-7B), 32.5 (CH$_2$, C-7A), 23.3 (CH$_2$, CH$_2$-8A), 14.3 (CH$_3$, C-9A). HRMS (ESI-) *m/z* 307.1707 [M-H]$^-$ (calcd for C$_{21}$H$_{23}$O$_2$, 307.1698).

3,3′-Diallyl-5-(2-hydroxyethyl)-(1,1′-biphenyl)-2,4′-diol (14c). Yield: 48% (51.1 mg). R$_f$ 0.42 (cyclohexane:acetone 65:35). ^{1}H-NMR ((CD$_3$)$_2$CO): δ 7.15 (bs, 1 H, H-6A), 7.12 (bd, *J* = 8.2 Hz, 1 H, H-6B), 6.91, (bs, 1 H, H-4A), 6.90 (bs, 1 H, H-2B), 6.88 (d, *J* = 8.2 Hz, 1 H, H-5B), 6.00 (m, 2 H, H-11A and H-8B), 5.09 (m, 2 H, CH$_2$-12A), 5.04 (m, 2 H, CH$_2$-9B), 3.80 (t, *J* = 6.6 Hz, 2 H, CH$_2$-8A), 3.40 (bt, *J* = 7.1 Hz, 4 H, CH$_2$-10A and CH$_2$-7B), 2.77 (t, *J* = 6.6 Hz, 2 H, CH$_2$-7A). ^{13}C-NMR ((CD$_3$)$_2$CO): δ 154.0 (C, C-4B), 148.8 (C, C-2A), 136.5 (CH, C-11A), 136.1 (CH, C-8B), 130.7 (CH, C-6A), 129.7 (C, C-5A), 129.4 (CH, C-4A), 128.7 (C, C-3A), 128.5 (CH, C-2B), 128.0 (CH, C-6B), 127.9 (C, C-3B), 126.8 (C,

C-1A), 126.1 (C, C-1B), 115.88 (CH$_2$, C-12A), 115.86 (CH$_2$, C-9B), 115.5 (CH, C-5B), 63.5 (CH$_2$, C-8A), 38.1 (CH$_2$, C-7A), 34.5 (CH$_2$, C-10A), 34.3 (CH$_2$, C-7B). HRMS (ESI-) *m/z* 309.1484 [M-H]$^-$ (calcd for C$_{20}$H$_{21}$O$_3$, 309.1491).

2-Allyl-4'-(allyloxy)-4-methoxy-6-propyl-(1,1'-biphenyl)-3-ol (15a). Yield: 15% (17.7 mg). R$_f$ 0.52 (cyclohexane:acetone 70:30). ^{1}H-NMR (CDCl$_3$): δ 7.03 (d, *J* = 7.5 Hz, 2 H, H-2B/H-6B), 6.92 (d, *J* = 7.5 Hz, 2 H, H-3B/H-5B), 6.66 (s, 1 H, H-5A), 6.11 (m, 1 H, H-8B), 5.82 (m, 1 H, H-11A), 5.59 (bs, 1 H, O*H*), 5.46 (dd, *J* = 17.2, 1.2 Hz, 1 H, C*H*$_a$-9B), 5.31 (dd, *J* = 10.5, 1.2 Hz, 1 H, C*H*$_b$-9B), 4.87 (dd, *J* = 10.1, 1.4 Hz, 1 H, C*H*$_a$-12A), 4.76 (dd, *J* = 17.0, 1.4 Hz, 1 H, C*H*$_b$-12A), 4.58 (d, *J* = 7.5 Hz, 2 H, C*H*$_2$-7B), 3.91 (s, 3 H, OC*H*$_3$), 3.12 (d, *J* = 6.1 Hz, 2 H, C*H*$_2$-10A), 2.24 (m, 2 H, C*H*$_2$-7A), 1.40 (m, 2 H, C*H*$_2$-8A), 0.76 (t, *J* = 7.5 Hz, 3 H, C*H*$_3$-9A). ^{13}C-NMR (CDCl$_3$): δ 157.5 (C, C-4B), 145.4 (C, COCH$_3$), 141.5 (C, C-3A), 136.9 (CH, C-11A), 134.6 (C, C-1A), 133.6 (CH, C-8B), 132.7 (C, C-6A), 132.6 (C, C-1B), 131.3 (CH, C-2B/C-6B), 124.6 (C, C-2A), 117.8 (CH$_2$, C-9B), 114.6 (CH$_2$, C-12A), 114.1 (CH, C-3B/C-5B), 109.3 (CH, C-5A), 69.0 (CH$_2$, C-7B), 56.1 (CH$_3$, OCH$_3$), 35.9 (CH$_2$, C-7A), 32.1 (CH$_2$, C-10A), 24.9 (CH$_2$, C-8A), 14.2 (CH$_3$, C-9A). HRMS (ESI-) *m/z* 337.1812 [M-H]$^-$ (calcd for C$_{22}$H$_{25}$O$_3$, 337.1804).

3.9. Human Cell Cultures

Two human colorectal adenocarcinoma cell lines (HCT-116 and HT-29) and one human prostate adenocarcinoma cell line (PC3) have been used in the present work and been cultured as previously reported [28,49]. The HCT-116 (ATCC number: CCL-247) cell line was cultured in McCoy's 5A (Medium, GlutaMAX™ Supplement, cod. 36600021 containing 3 g/L of D-glucose), HT-29 (ATCC number: HTB-38) was cultured in in DMEM medium (Dulbecco's Modified Eagle Medium 1X; GIBCO, Cat No. 31965-023 containing 4.5g/L of D-glucose) while PC3 cell line (ATCC number: CRL-1435) was cultured in DMEM/F-12, (Dulbecco's Modified Eagle Medium/Nutrient Mixture F-12, GlutaMAX™ Supplement, cod. 31331028). Each medium was supplemented with 10% Fetal Bovine Serum (FBS) and 100 U/ml of penicillin-streptomycin. The cell cultures were incubated at 37 °C in humidified atmosphere with 5% of CO$_2$ and 95% of air. The medium was changed twice in the week.

3.10. Antiproliferative Assay

HCT-116 HT-29 and PC3 cell lines (1.8 and 2.5 × 10^3 cells/0.33 cm^2 respectively) were plated in 96 well plates from "Nunclon™ Microwell™" (Thermo Fisher Scientific, Bremen, Germany), and were incubated at 37 °C. After 24 h cells (60% confluence) were treated with compounds 2-9, 12a-c, 13a-c, 14a-c, and 15a at increasing concentration from 0.001 to 100 µM. The cellular vitality and/or the cellular cytotoxicity has been evaluated by colorimetric assay with tetrazolium salt MTT (3-(4,5-methylthiazol-2-yl)-2,5-diphenyltetrazolium bromide) as previously reported [27]. After 72 hours of treatment with the bisphenol neolignans, MTT salt was added and maintained for 3 hours. The purple formazan dye, produced by the metabolic process of vital cells, was solubilized by adding 150 µL/well of DMSO (dimethyl sulfoxide) and the optical density (OD) values were read on a multiwell scanning spectrophotometer (Multiskan™ reader, Thermo Fisher Scientific, Bremen, Germany), using a wavelength of 570 nm. Each value was the average of four–six wells. The percentage of cellular growth was calculated according to NCI [50]: 100 × (T − T$_0$) / (C−T$_0$) (where T is the optical density of the test well after a 72 h period of exposure to test compound, T$_0$ is the optical density at time zero, and C is the optical density of the untreated control cell cultures). When the optical density of treated cells was lower than the T$_0$ value the following formula was used: 100 × (T − T$_0$) /T$_0$. GI$_{50}$ values were calculated by the software GraphPad Prism v.6.

3.11. Apoptosis Analysis by Imaging Flow Cytometry

HCT116 and PC3 cell lines were cultured in six-well plates (about 3.5 × 10^5 cells/9.6 cm^2) and treated with the following compounds: 14a and c, and 15a. Vehicle-treated cells and the anticancer 5-fluorouracil (5-FU, final concentration of 10 µM) were used as negative and positive control, respectively. Cells were harvested by trypsinization after 72 h of treatment, washed in PBS, and stained with Alexa Fluor®488 dye conjugated to annexin V and propidium iodide (PI) in according to manufacturing protocol (The Alexa Fluor®488 annexin V /Dead Cell Apoptosis Kit by Invitrogen, Cat. V13245; Thermo Fisher Scientific, Bremen, Germany). PI is impermeable to live cells and to cells in early apoptosis, but stains dead cells and late apoptosis cells with red fluorescence, binding tightly to the nucleic acids while Alexa Fluor®488 dye, conjugated to annexin V, stains in green the apoptotic cells. After staining the cellular suspension was analyzed on a flow cytometer (Amnis FlowSight Millipore, Merck KgaA, Darmstadt, Germany), and results were analyzed using Image Data Exploration and Analysis (IDEAS) software (Amnis part of EMD Millipore, Seattle, WA, USA).

4. Conclusions

The present work reports the synthesis of a library of bisphenol neolignans inspired by honokiol. These products were evaluated for their potential antiproliferative activity against three cancer cell lines, namely HCT-116, HT-29 (both human colorectal adenocarcinoma) and PC3 (human prostate adenocarcinoma). This study suggests some structural motifs important for the activity and overall highlights that bisphenol neolignans 9, 12b, 14a-c, and 15a, structurally related to honokiol, show higher antiproliferative activity than the natural lead. The most promising antiproliferative agents (14a, 14c, and 15a) were selected for a flow cytometric analysis; this showed that the inhibition of proliferation is mainly due to apoptotic process. In conclusion, some of the bisphenol neolignans reported in the present study have shown promising properties for the development of new natural-derived antitumor agents.

Supplementary Materials: A copy of HRMS and NMR spectra of the new synthesized bisphenol neolignans is available online at http://www.mdpi.com/1420-3049/25/3/733/s1.

Author Contributions: Conceptualization, C.T. and D.F.C.; Methodology, N.C., V.M. and V.B.; Validation, V.M., N.C., V.B., and G.S.; Investigation, N.C., G.S., and M.D.; Resources, V.M., C.T. and D.F.C.; Data curation, N.C., V.M. and V.B.; Writing—Original draft preparation, N.C., V.M. and V.B.; Writing—Review and editing, C.T., D.F.C. and V.B.; Visualization, N.C., G.S., and V.B.; Supervision, N.C., V.M., and V.B.; Project administration, C.T.; Funding acquisition, V.M. and V.B.

Funding: This research was funded by 'Piano della Ricerca di Ateneo 2016-2018, Linea d'intervento 2' of Università degli Studi di Catania, by MIUR ITALY PRIN 2017 (Project No. 2017A95NCJ), and partially supported by project "FIR 2014, University of Catania, Italy" (project No. 668A01).

Acknowledgments: The authors acknowledge the PON project Bio-nanotech Research and Innovation Tower (BRIT), financed by the Italian Ministry for Education, University and Research (MIUR) (Grant no. PONa3_00136).

Conflicts of Interest: The authors declare no conflict of interest. The funders had no role in the design of the study; in the collection, analyses, or interpretation of data; in the writing of the manuscript, or in the decision to publish the results.

References

1. Zhang, J.; Chen, J.J.; Liang, Z.Z.; Zhao, C.Q. New lignans and their biological activities. *Chem. Biodivers.* **2014**, *11*, 1–54. [CrossRef] [PubMed]
2. Aldemir, H.; Richarz, R.; Gulder, T.A.M. The biocatalytic repertoire of natural biaryl formation. *Ange. Chem. Int. Ed.* **2014**, *53*, 8286–8293. [CrossRef] [PubMed]
3. Lee, Y.J.; Lee, Y.M.; Lee, C.K.; Jung, J.K.; Han, S.B.; Hong, J.T. Therapeutic applications of compounds in the Magnolia family. *Pharmacol. Ther.* **2011**, *130*, 157–176. [CrossRef] [PubMed]
4. Patočka, J.; Jakl, J.; Strunecká, A. Expectations of biologically active compounds of the genus *Magnolia* in biomedicine. *J. Appl. Biomed.* **2006**, *4*, 171–178. [CrossRef]
5. Kelm, M.A.; Nair, M.G. A brief summary of biologically active compounds from *Magnolia* spp. *Stud. Nat. Prod. Chem.* **2000**, *24*, 845–873.
6. Shen, J.L.; Man, K.M.; Huang, P.H.; Chen, W.C.; Chen, D.C.; Cheng, Y.W.; Liu, P.L.; Chou, M.C.; Chen, Y.H. Honokiol and magnolol as multifunctional antioxidative molecules for dermatologic disorders. *Molecules* **2010**, *15*, 6452–6465. [CrossRef]

7. Amorati, R.; Zotova, J.; Baschieri, A.; Valgimigli, L. Antioxidant activity of magnolol and honokiol: Kinetic and mechanistic investigations of their reaction with peroxyl radicals. *J. Org. Chem.* **2015**, *80*, 10651–10659. [CrossRef]
8. Lin, Y.R.; Chen, H.H.; Ko, C.H.; Chan, M.H. Effects of honokiol and magnolol on acute and inflammatory pain models in mice. *Life Sci.* **2007**, *81*, 1071–1078. [CrossRef]
9. Lin, Y.R.; Chen, H.H.; Ko, C.H.; Chan, M.H. Neuroprotective activity of honokiol and magnolol in cerebellar granule cell damage. *Eur. J. Pharmacol.* **2006**, *537*, 64–69. [CrossRef]
10. Fried, L.E.; Arbiser, J.L. Honokiol, a multifunctional antiangiogenic and antitumor agent. *Antioxid. Redox Signal.* **2009**, *11*, 1139–1148. [CrossRef]
11. Ranaware, A.M.; Banik, K.; Deshpande, V.; Padmavathi, G.; Roy, N.K.; Sethi, G.; Fan, L.; Kumar, A.P.; Kunnumakkara, A.B. Magnolol: A neolignan from the Magnolia family for the prevention and treatment of cancer. *Int. J. Mol. Sci.* **2018**, *19*, 2362. [CrossRef]
12. Kim, G.D.; Oh, J.; Park, H.J.; Bae, K.; Lee, S.K. Magnolol inhibits angiogenesis by regulating ROS–Mediated apoptosis and the PI3K/AKT/mTOR signaling pathway in mES/EB-derived endothelial-like cells. *Int. J. Oncol.* **2013**, *43*, 600–610. [CrossRef] [PubMed]
13. Shen, J.; Ma, H.L.; Zhang, T.C.; Liu, H.; Yu, L.H.; Li, G.S.; Li, H.S.; Hu, M.C. Magnolol inhibits the growth of non-small cell lung cancer via inhibiting microtubule polymerization. *Cell. Physiol. Biochem.* **2017**, *42*, 1789–1801. [CrossRef] [PubMed]
14. Shen, L.; Zhang, F.; Huang, R.M.; Yan, J.; Shen, B. Honokiol inhibits bladder cancer cell invasion through repressing SRC-3 expression and epithelial-mesenchymal transition. *Oncol. Lett.* **2017**, *14*, 4294–4300. [CrossRef] [PubMed]
15. Yeh, P.S.; Wang, W.; Chang, Y.A.; Lin, C.J.; Wang, J.J.; Chen, R.M. Honokiol induces autophagy of neuroblastoma cells through activating the PI3K/Akt/mTOR and endoplasmic reticular stress/ERK1/2 signaling pathways and suppressing cell migration. *Cancer Lett.* **2016**, *370*, 66–77. [CrossRef]
16. Hajduk, P.J.; Bures, M.; Praestgaard, J.; Fesik, S.W. Privileged molecules for protein binding identified from NMR-based screening. *J. Med. Chem.* **2000**, *43*, 3443–3447. [CrossRef]
17. Jada, S.; Doma, M.R.; Singh, P.P.; Kumar, S.; Malik, F.; Sharma, A.; Khan, I.A.; Qazi, G.N.; Kumar, H.M.S. Design and synthesis of novel magnolol derivatives as potential antimicrobial and antiproliferative compounds. *Eur. J. Med. Chem.* **2012**, *51*, 35–41. [CrossRef]
18. Amblard, F.; Govindarajan, B.; Lefkove, B.; Rapp, K.L.; Detorio, M.; Arbiser, J.L.; Schinazi, R.F. Synthesis, cytotoxicity, and antiviral activities of new neolignans related to honokiol and magnolol. *Bioorg. Med. Chem. Lett.* **2007**, *17*, 4428–4431. [CrossRef]
19. Tripathi, S.; Chan, M.H.; Chen, C.P. An expedient synthesis of honokiol and its analogues as potential neuropreventive agents. *Bioorg. Med. Chem. Lett.* **2012**, *22*, 216–221. [CrossRef]
20. Lee, S.H.; Fei, X.; Lee, C.; Do, H.T.T.; Rhee, I.; Seo, S.Y. Synthesis of either C2-or C4′-alkylated derivatives of honokiol and their biological evaluation for anti-inflammatory activity. *Chem. Pharm. Bull.* **2019**, *67*, 966–976. [CrossRef]
21. Lin, J.M.; Gowda, A.S.P.; Sharma, A.K.; Amin, S. In vitro growth inhibition of human cancer cells by novel honokiol analogs. *Bioorg. Med. Chem.* **2012**, *20*, 3202–3211. [CrossRef] [PubMed]
22. Maioli, M.; Basoli, V.; Carta, P.; Fabbri, D.; Dettori, M.A.; Cruciani, S.; Serra, P.A.; Delogu, G. Synthesis of magnolol and honokiol derivatives and their effect against hepatocarcinoma cells. *PLoS ONE* **2018**, *13*, e0192178. [CrossRef] [PubMed]
23. Sanchez-Peris, M.; Murga, J.; Falomir, E.; Carda, M.; Marco, J.A. Synthesis of honokiol analogues and evaluation of their modulating action on VEGF protein secretion and telomerase-related gene expressions. *Chem. Biol. Drug Des.* **2017**, *89*, 577–584. [CrossRef] [PubMed]
24. Kotha, S.; Lahiri, K.; Kashinath, D. Recent applications of the Suzuki–Miyaura cross-coupling reaction in organic synthesis. *Tetrahedron* **2002**, *58*, 9633–9695. [CrossRef]
25. Martin, R.; Buchwald, S.L. Palladium-catalyzed Suzuki–Miyaura cross-coupling reactions employing dialkylbiaryl phosphine ligands. *Acc. Chem. Res.* **2008**, *41*, 1461–1473. [CrossRef]
26. Pulvirenti, L.; Muccilli, V.; Cardullo, N.; Spatafora, C.; Tringali, C. Chemoenzymatic synthesis and alpha-glucosidase inhibitory activity of dimeric neolignans inspired by magnolol. *J. Nat. Prod.* **2017**, *80*, 1648–1657. [CrossRef]

27. Cardullo, N.; Spatafora, C.; Musso, N.; Barresi, V.; Condorelli, D.; Tringali, C. Resveratrol-related polymethoxystilbene glycosides: Synthesis, antiproliferative activity, and glycosidase inhibition. *J. Nat. Prod.* **2015**, *78*, 2675–2683. [CrossRef]
28. Cardullo, N.; Pulvirenti, L.; Spatafora, C.; Musso, N.; Barresi, V.; Condorelli, D.F.; Tringalii, C. Dihydrobenzofuran neolignanamides: Laccase-mediated biomimetic synthesis and antiproliferative activity. *J. Nat. Prod.* **2016**, *79*, 2122–2134. [CrossRef]
29. Chillemi, R.; Cardullo, N.; Greco, V.; Malfa, G.; Tomasello, B.; Sciuto, S. Synthesis of amphiphilic resveratrol lipoconjugates and evaluation of their anticancer activity towards neuroblastoma SH-SY5Y cell line. *Eur. J. Med. Chem.* **2015**, *96*, 467–481. [CrossRef]
30. Capolupo, A.; Tosco, A.; Mozzicafreddo, M.; Tringali, C.; Cardullo, N.; Monti, M.C.; Casapullo, A. Proteasome as a new target for bio-inspired benzo[k,l]xanthene lignans. *Chem. Eur. J.* **2017**, *23*, 8371–8374. [CrossRef]
31. Spatafora, C.; Barresi, V.; Bhusainahalli, V.M.; Di Micco, S.; Musso, N.; Riccio, R.; Bifulco, G.; Condorelli, D.; Tringali, C. Bio-inspired benzo[k,l]xanthene lignans: Synthesis, DNA-interaction and antiproliferative properties. *Org. Biomol. Chem.* **2014**, *12*, 2686–2701. [CrossRef]
32. Cardullo, N.; Catinella, C.; Floresta, G.; Muccilli, V.; Rosselli, S.; Rescifina, A.; Bruno, M.; Tringali, C. Synthesis of rosmarinic acid amides as antioxidative and hypoglycemic agents. *J. Nat. Prod.* **2019**, *82*, 573–582. [CrossRef]
33. Cardullo, N.; Muccilli, V.; Pulvirenti, L.; Cornu, A.; Poiségu, L.; Deffieux, D.; Quideau, S.; Tringali, C. C-glucosidic ellagitannins and galloylated glucoses as potential functional food ingredients with anti-diabetic properties: A study of α-glucosidase and α-amylase inhibition. *Food Chem.* **2020**, *313*, 126099. [CrossRef] [PubMed]
34. Genovese, C.; Pulvirenti, L.; Cardullo, N.; Muccilli, V.; Tempera, G.; Nicolosi, D.; Tringali, C. Bioinspired benzoxanthene lignans as a new class of antimycotic agents: Synthesis and *Candida* spp. growth inhibition. *Nat. Prod. Res.* **2018**. publicated ahead of print. [CrossRef] [PubMed]
35. Di Micco, S.; Spatafora, C.; Cardullo, N.; Riccio, R.; Fischer, K.; Pergola, C.; Koeberle, A.; Werz, O.; Chalal, M.; Vervandier-Fasseur, D.; et al. 2,3-Dihydrobenzofuran privileged structures as new bioinspired lead compounds for the design of mPGES-1 inhibitors. *Bioorg. Med. Chem.* **2016**, *24*, 820–826. [CrossRef] [PubMed]
36. Gerstmeier, J.; Kretzer, C.; Di Micco, S.; Miek, L.; Butschek, H.; Cantone, V.; Bilancia, R.; Rizza, R.; Troisi, F.; Cardullo, N.; et al. Novel benzoxanthene lignans that favorably modulate lipid mediator biosynthesis: A promising pharmacological strategy for anti-inflammatory therapy. *Biochem. Pharmacol.* **2019**, *165*, 263–274. [CrossRef] [PubMed]
37. Di Micco, S.; Pulvirenti, L.; Bruno, I.; Terracciano, S.; Russo, A.; Vaccaro, M.C.; Ruggiero, D.; Muccilli, V.; Cardullo, N.; Tringali, C.; et al. Identification by inverse virtual screening of magnolol-based scaffold as new tankyrase-2 inhibitors. *Bioorg. Med. Chem.* **2018**, *26*, 3953–3957. [CrossRef] [PubMed]
38. Baschieri, A.; Pulvirenti, L.; Muccilli, V.; Amorati, R.; Tringali, C. Chain-breaking antioxidant activity of hydroxylated and methoxylated magnolol derivatives: The role of H-bonds. *Org. Biomol. Chem.* **2017**, *15*, 6177–6184. [CrossRef]
39. Cardile, V.; Lombardo, L.; Spatafora, C.; Tringali, C. Chemo-enzymatic synthesis and cell-growth inhibition activity of resveratrol analogues. *Bioorg. Chem.* **2005**, *33*, 22–33. [CrossRef]
40. Ng, S.Y.; Cardullo, N.; Yeo, S.C.M.; Spatafora, C.; Tringali, C.; Ong, P.-S.; Lin, H.-S. Quantification of the resveratrol analogs trans-2,3-dimethoxy-stilbene and trans-3,4-dimethoxystilbene in rat plasma: Application to pre-clinical pharmacokinetic studies. *Molecules* **2014**, *19*, 9577–9590. [CrossRef]
41. Schuhly, W.; Hufner, A.; Pferschy-Wenzig, E.M.; Prettner, E.; Adams, M.; Bodensieck, A.; Kunert, O.; Oluwemimo, A.; Haslinger, E.; Bauer, R. Design and synthesis of ten biphenyl-neolignan derivatives and their in vitro inhibitory potency against cyclooxygenase-1/2 activity and 5-lipoxygenase-mediated LTB4-formation. *Bioorg. Med. Chem.* **2009**, *17*, 4459–4465. [CrossRef] [PubMed]
42. Hu, Y.; Shen, Y.F.; Tu, X.; Wu, X.H.; Wang, G.X.; Ling, F. Isolation of anti-*Saprolegnia* lignans from *Magnolia officinalis* and SAR evaluation of honokiol/magnolol analogs. *Bioorg. Med. Chem. Lett.* **2019**, *29*, 389–395. [CrossRef] [PubMed]
43. Tanaka, T.; Sakurai, Y.; Okazaki, H.; Hasegawa, T.; Fukuyama, Y. Biphenyl Derivative Composition for Nerve Cell Degeneration Repairing or Protective Agent and Process for Preparing a Phenyl Derivative Contained in the Composition. U.S. Patent 5053548, 1 October 1991.

44. Bovicelli, P.; Antonioletti, R.; Mancini, S.; Causio, S.; Borioni, G.; Annnendola, S.; Barontini, M. Expedient synthesis of hydroxytyrosol and its esters. *Synt. Commun.* **2007**, *37*, 4245–4252. [CrossRef]
45. Ichikawa, T.; Netsu, M.; Mizuno, M.; Mizusaki, T.; Takagi, Y.; Sawama, Y.; Monguchi, Y.; Sajiki, H. Development of a unique heterogeneous palladium catalyst for the Suzuki–Miyaura reaction using (hetero)aryl chlorides and chemoselective hydrogenation. *Adv. Synth. Catal.* **2017**, *359*, 2269–2279. [CrossRef]
46. Brunel, J.M. Scope, limitations and mechanistic aspects in the selective homogeneous palladium-catalyzed reduction of alkenes under transfer hydrogen conditions. *Tetrahedron* **2007**, *63*, 3899–3906. [CrossRef]
47. Freudenberg, K.; Renner, K.C. Biphenyls and diaryl ethers among the precursors of lignin. *Chem. Ber.* **1965**, *98*, 1879–1892. [CrossRef]
48. Zhang, J.; Sun, T.J.; Diao, H.P.; Li, M.P. Quantitative structure-activity relationship studies of phenol's biological activity. *Shanxi Daxue Xuebao* **2011**, *34*, 468–474.
49. Accardo, A.; Del Zoppo, L.; Morelli, G.; Condorelli, D.F.; Barresi, V.; Musso, N.; Spampinato, G.; Bellia, F.; Tabbi, G.; Rizzarelli, E. Liposome antibody-ionophore conjugate antiproliferative activity increases by cellular metallostasis alteration. *Med. Chem. Comm.* **2016**, *7*, 2364–2367. [CrossRef]
50. Scherf, U.; Ross, D.T.; Waltham, M.; Smith, L.H.; Lee, J.K.; Tanabe, L.; Kohn, K.W.; Reinhold, W.C.; Myers, T.G.; Andrews, D.T.; et al. A gene expression database for the molecular pharmacology of cancer. *Nat. Genet.* **2000**, *24*, 236–244. [CrossRef]

Sample Availability: Samples of the compounds 3-9, 12a-c, 13a-c, 14a-c and 15a are available from the authors.

 © 2020 by the authors. Licensee MDPI, Basel, Switzerland. This article is an open access article distributed under the terms and conditions of the Creative Commons Attribution (CC BY) license (http://creativecommons.org/licenses/by/4.0/).

Article

Multicomponent Synthesis of Polyphenols and Their In Vitro Evaluation as Potential β-Amyloid Aggregation Inhibitors

Denise Galante [1], Luca Banfi [2], Giulia Baruzzo [2], Andrea Basso [2], Cristina D'Arrigo [1], Dario Lunaccio [1], Lisa Moni [2], Renata Riva [3] and Chiara Lambruschini [2,*]

[1] Istituto per lo Studio delle Macromolecole, Consiglio Nazionale delle Ricerche, via De Marini 6, 16149 Genova, Italy
[2] Department of Chemistry and Industrial Chemistry, University of Genova, via Dodecaneso 31, 16146 Genova, Italy
[3] Department of Pharmacy, Università di Genova, viale Cembrano 4, 16147 Genova, Italy
* Correspondence: chiara.lambruschini@unige.it; Tel.: +39-010-3536119; Fax: +39-010-3538733

Academic Editor: Corrado Tringali
Received: 27 June 2019; Accepted: 18 July 2019; Published: 19 July 2019

Abstract: While plant polyphenols possess a variety of biological properties, exploration of chemical diversity around them is still problematic. Here, an example of application of the Ugi multicomponent reaction to the combinatorial assembly of artificial, yet "natural-like", polyphenols is presented. The synthesized compounds represent a second-generation library directed to the inhibition of β-amyloid protein aggregation. Chiral enantiopure compounds, and polyphenol-β-lactam hybrids have been prepared too. The biochemical assays have highlighted the importance of the key pharmacophores in these compounds. A lead for inhibition of aggregation of truncated protein AβpE3-42 was selected.

Keywords: multicomponent reactions; polyphenols; β-amyloid proteins; Alzheimer's disease; Ugi reaction; β-lactams

1. Introduction

Natural polyphenols of plant origin are important elements of our diet and, for this reason, their biological properties have been thoroughly studied [1,2]. Their most renowned characteristic is their antioxidant activity, which is believed to play an important role in preventing age-related diseases such as atherosclerosis. Since long ago, plant polyphenols have attracted much interest in the nutraceutical and cosmetic fields. On the other hand, the pharmaceutical industry has remained so far rather detached regarding the investigation of plant polyphenols as possible leads for drug development [3]. The reasons for this lack of interest are not fully clear but may be related to the poor pharmacokinetic properties of natural polyphenols, to their in vivo instability, and to the challenging synthetic modification of the natural members of this family. Actually, only few studies on the preparation of artificial analogues through total synthesis or semi-synthesis have been reported so far [3–12]. Nevertheless, several reports have pointed out that polyphenols, apart from anti-oxidant properties, may have a variety of other biological effects, such as anti-inflammatory [13], anti-cancer [12,14,15], anti-microbial [16], and anti-hyperglycemic [17] activity. Last but not least, some natural polyphenols have been demonstrated to be able to inhibit β-amyloid aggregation, thus being promising for the prevention of the Alzheimer's disease [18–21].

We reasoned that a "natural fragment-based approach" [22] to the combinatorial synthesis of artificial polyphenols starting from small, phenolic, building blocks, would allow a more systematic exploration of their chemical space, allowing to select possible hits for drug discovery, endowed with modulable pharmacodynamic and pharmacokinetic properties. In particular, the Ugi multi-component reaction seemed particularly well suited for this approach, since it leads to mixed polyphenol-peptidomimetic structures and allows rigidification through post-MCR cyclization steps [23,24]. A similar approach was recently used by Ismaili et al. [7], who developed a multi-target lead for Alzheimer's disease by joining ferulic acid to an acetylcholine esterase inhibitor, lipoic acid and a melatonin analogue. However, in Ismaili's work, only one of the four components used as inputs in the Ugi reaction was a phenol, and the final product was not indeed polyphenols. On the contrary, in our plan, depicted in Figure 1, up to four of the components would be of phenolic nature, thus leading to true artificial, polyphenols of general formula **1**. Our preliminary results using this strategy were reported in a previous paper [25]. Now, also on the basis of the biochemical properties of the first-generation library, we have extended exploration of the chemical space of these polyphenols, preparing new chemical entities, also exploiting components that are chiral enantiopure or that allow post-MCR cyclization.

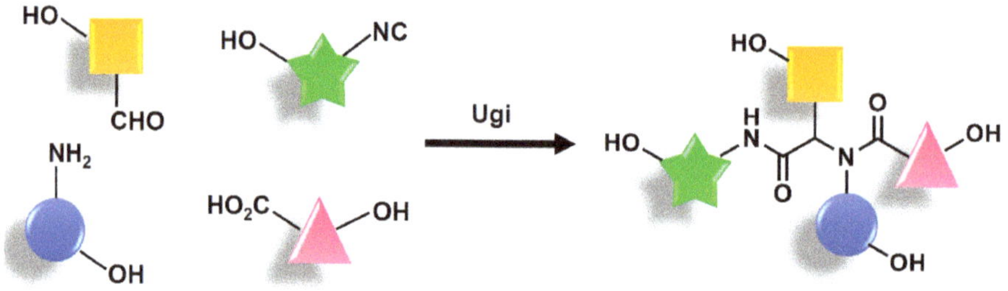

Figure 1. General strategy for assembling artificial polyphenols from simple phenolic building blocks through the Ugi reaction.

2. Results and Discussion

2.1. Synthesis

Scheme 1 shows a specific example of the general synthetic strategy used for the preparation of polyphenols. This strategy was optimized also on the basis of the outcomes of our initial preliminary work. The need, in particular for in vivo experiments, of high purity final polyphenols **8**, prompted us to prepare compounds **8** through a high yielding solvolysis of polyacetates **7**. In this way, the purification is very simple, and we can avoid an extractive and chromatographic purification of the polyphenols, which, in some cases, although not always, gave insufficiently pure compounds. On the other hand, the acetate protecting group was found to be not fully stable under the conditions of the Ugi reaction. Thus, our standard, optimized, procedure involved the use of phenolic building blocks protected as allyl ethers. The Ugi reaction gave best results when preformation of the imine was implemented and CF_3CH_2OH (TFE)/ethanol mixture was used as the solvent. Removal of the allyl protecting groups was best performed with a minimum amount of Pd catalyst and ammonium formate as the scavenger, followed by immediate acetylation. Chromatographic purification afforded acetates **7**, which were obtained in high purity and fully characterized at this level. Finally, deacetylation was performed with a protocol that completely avoided both extractive and chromatographic purification of the final polyphenols **8**. As shown in Scheme 1, for compound **8c** this procedure was very efficient, and the final HPLC purity was 99%.

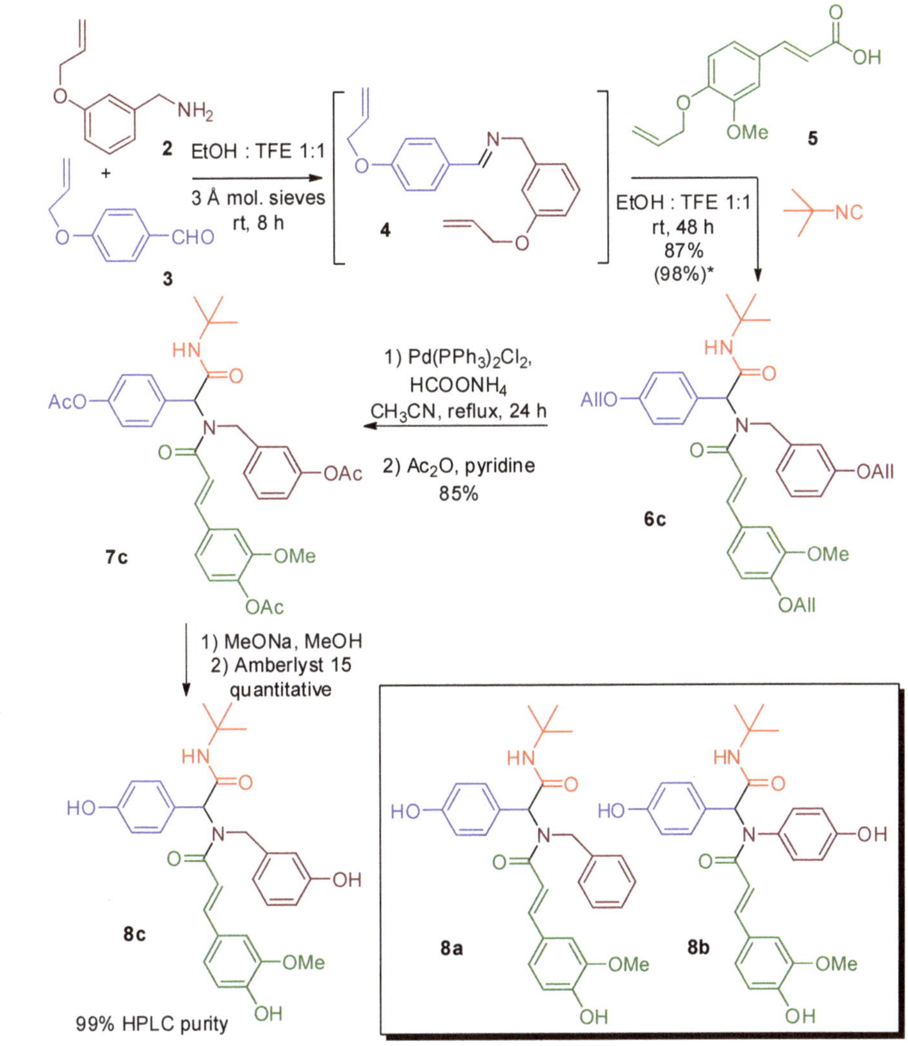

*Yield calculated taking into account the recovered aldehyde

Scheme 1. Representative example of the synthesis of artificial polyphenols.

Other similar polyphenols prepared by us are shown in Figure 2. In designing the new entities to be synthesized, we took as a model the best compounds that emerged from the preliminary studies [25]. In that paper, we tested the polyphenols on two amyloid proteins: Aβ1-42 and AβpE3-42. For the first one, the best compound was found to be **8a**. On the smaller fragment, the best hits contained a residue derived from caffeic acid, but we later found that polyphenols containing the catechol moiety typical of caffeic acid where less stable and, most of all, cytotoxic to neuronal cells [26]. Thus, we decided to keep, as the lead for AβpE3-42, compound **8b**, which was the best among those derived from ferulic acid.

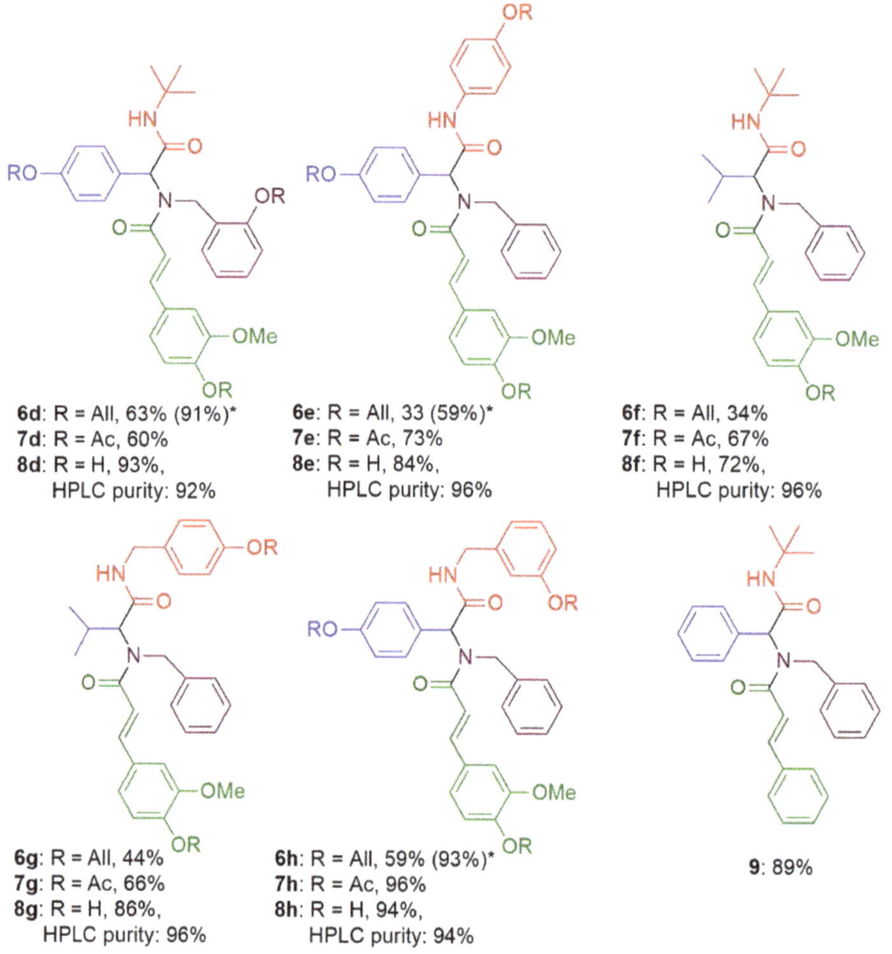

6d: R = All, 63% (91%)*
7d: R = Ac, 60%
8d: R = H, 93%,
   HPLC purity: 92%

6e: R = All, 33 (59%)*
7e: R = Ac, 73%
8e: R = H, 84%,
   HPLC purity: 96%

6f: R = All, 34%
7f: R = Ac, 67%
8f: R = H, 72%,
   HPLC purity: 96%

6g: R = All, 44%
7g: R = Ac, 66%
8g: R = H, 86%,
   HPLC purity: 96%

6h: R = All, 59% (93%)*
7h: R = Ac, 96%
8h: R = H, 94%,
   HPLC purity: 94%

9: 89%

* Yield calculated taking into account the recovered aldehyde

Figure 2. Simple analogues of **8a** prepared.

Since the ferulic derived part seemed to be very important, in this study we maintained the carboxylic building block, and varied only the other three components. For example, looking at the structure of **8a**, we replaced benzylamine with a phenol containing benzylamine (see **8c** and **8d**). In **8f** and **8g** we replaced *p*-hydroxybenzaldehyde (that was the aldehyde component for preparation of both **8a** and **8b**) with an aliphatic aldehyde (isobutyraldehyde). In **8e** we replaced *t*-butyl isocyanide with a phenol containing an aromatic isocyanide, whereas for **8g** and **8h** a benzyl isocyanide containing a phenol was used. Finally, we also prepared compound **9**, which is the analogue of **8a**, totally devoid of phenolic groups, in order to check the importance of this moiety for biological activity.

As far as it concerns the overall yields of the various syntheses, the most critical step was found to be the Ugi reaction. In general, we can say that aromatic isocyanides (see **6d**) and aliphatic aldehydes (see **6e** and **6f**) bring about a less efficient multicomponent reaction. The high yield achieved for compound **9** demonstrates that also protected ferulic acid, as the carboxylic component, is not ideal for the Ugi reaction. The decrease in the yield is typically due to sluggish reactions and incomplete conversion. In some instances, we have determined the yield considering the recovered starting aldehyde. However, it should be noted that, since our main goal was to assess the biochemical properties of our polyphenols, the individual syntheses have not been optimized.

All the polyphenols prepared in the previous paper and those depicted in Scheme 1 and Figure 2 are racemic. To verify the possible influence of the absolute configuration of the stereogenic center generated during the Ugi reaction, we decided to use, as amine component, enantiopure α-methylbenzylamine **10** (Scheme 2).

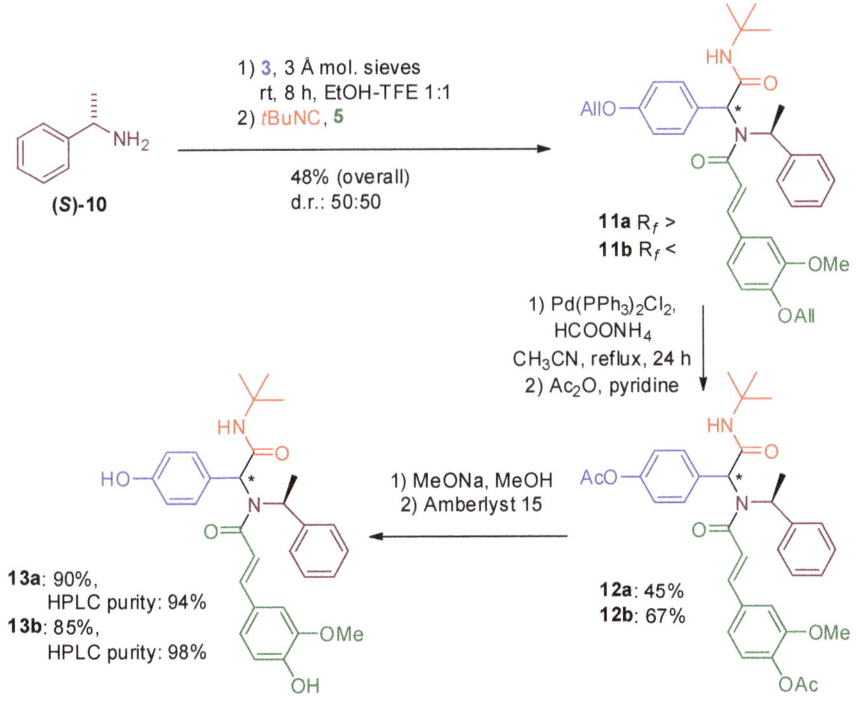

Scheme 2. Preparation of enantiopure polyphenols **13a,b**.

As expected, the Ugi reaction was poorly diastereoselective (and slower compared to the ones employing unsubstituted benzylamines). The two diastereomers could be conveniently separated at this stage, and then independently converted into enantiopure polyphenols **13a** and **13b**.

Finally, we wanted to prepare some rigidified analogues, by applying a post-MCR cyclization. In a previous work within our group [27], we have noticed that Ugi adducts derived from glycolaldehyde could be cyclized to β-lactams, exploiting the isocyanide derived secondary amide as the nucleophile in an intramolecular S_N2 process [23]. This interesting reaction was not fully explored, but we reasoned that in the present case it could offer a possibility to obtain polyphenol-β-lactam hybrids as rigidified analogues of **8a** and **8b**. Scheme 3 shows this approach.

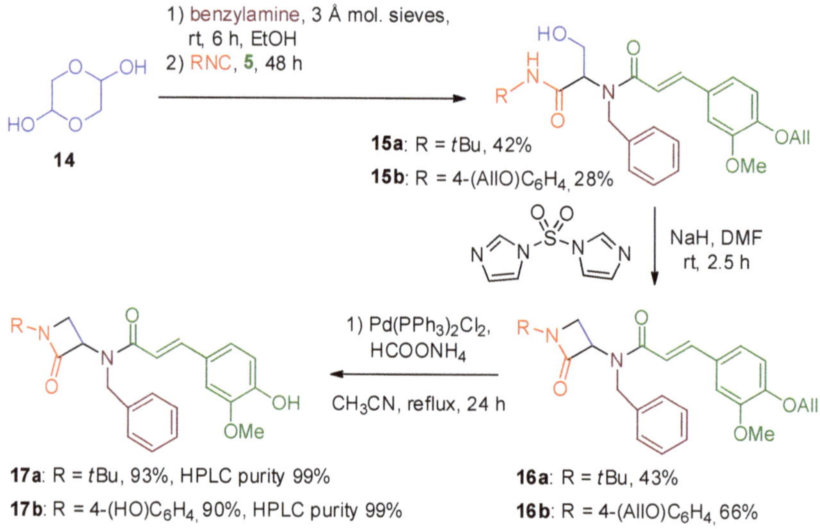

Scheme 3. Preparation of β-lactams **17a,b**.

Glycolaldehyde dimer **14** has been already used by us and others in Ugi or Ugi-type reactions [27–29]. Ugi reactions with **14** are known to proceed only in moderate yields, and, as stated above, protected ferulic acid, like other α,β-unsaturated acids, is not an ideal component for this MCR. Thus, we were not surprised by the low to moderate yields achieved for **15a** and **15b**. Cyclization was then carried out using sulfonyl diimidazole in the presence of NaH. This method, first introduced by Hanessian [30,31], is a variant of the better renown Mitsunobu reaction, and offers advantages in terms of operational simplicity and atom economy [32,33]. Formation of the β-lactams took place quite fast, but it was much cleaner starting with the more acidic aromatic secondary amine **16b**, whereas with the *t*-butyl derivative **16a** several side products were present, lowering the yields.

In this case, we preferred to avoid the final deacetylation. Therefore, after deprotection of the allyl groups, pure final phenols **17a,b** were directly obtained by chromatography in highly pure form, being not susceptible to degradation under chromatographic conditions.

2.2. Biochemical Assays

Then, we investigated the interaction of our candidates with two β-amyloid peptides, Aβ1-42 and AβpE3-42. While the full-length Aβ1-42 is one of the most abundantly identified in the brain deposits, AβpE3-42 is a peptide *N*-terminal truncated at residue 3 (Glu) and further modified by cyclization of Glu (E) to pyroglutamic acid (pE). These structural modifications are known to increase AβpE3-42 aggregation propensity [34].

First, in order to check that no precipitation of our phenols could occur under the assay conditions, we determined the solubility in phosphate buffer solution (PBS) at pH 7.4 containing 1% of DMSO. This percentage of DMSO does not alter the aggregation of β-amyloids, as demonstrated by control experiments carried out with the solvent alone. This analysis was performed by UV-VIS spectroscopic monitoring of the solutions of the compounds at different concentrations, both at the λ_{max} (at around 330 nm) and at 405 nm to evidence the formation of turbidity. Only compound **8e** was found to be fully soluble in the whole range of concentrations tested (10–500 µM). In this case the absorbance curve at 333 nm was linear and the line at 405 nm flat. For the other compounds bending of the absorbance curve and increase of turbidity started at the concentrations indicated in Table 1, which were anyway far beyond the one used for the subsequent experiments (25 µM). Interestingly, most of the newly synthesized compounds turned out to be more soluble than our previous lead **8a**, and this may represent ad advantage for in vivo tests. The aggregation inhibition experiments were carried out with the thioflavin T methodology. This assay was selected during our previous work as the most helpful in order to predict activity of our polyphenols. We also preliminary tried to investigate the activity of our polyphenols by ANS fluorescence. However, ANS tests were not informative in the case of compounds derived from ferulic acid, since they showed major spectral interferences with ANS. On the other hand, also circular dichroism experiments were not useful, because of interference of 1% DMSO, which was needed to keep our compounds in solution. During our preliminary work [25,26], we have also used electron microscopy and NMR studies, which confirmed the thioflavin data. However, these long studies were carried out only on **8a** or other few selected compounds, and are not suited for a fast selection of the here reported second generation compounds, that was made with thioflavin T.

Table 1. Solubility and Thioflavin assays on compounds **8a,g**, **9**, **13a,b**, **17a,b**.

Entry	Compound	Solubility [1]	Plateau Aβ1-42 [2]	Plateau AβpE3-42 [2]
1	control	-	100%	100%
2	8a	100 μM	50%	90%
3	8b	250 μM	65%	53%
4	8c	400 μM	74%	51%
5	8d	200 μM	68%	89%
6	8e	250 μM	68%	72%
7	8f	>500 μM	86%	192%
8	8g	100 μM	102%	127%
9	8h	200 μM	117%	72%
10	9	100 μM	136%	104%
11	13a	150 μM	87%	105%
12	13b	100 μM	86%	77%
13	17a	250 μM	96%	96%
14	17b	250 μM	70%	72%

[1] Concentration where the absorbance curve at about 330 nm starts to deflect from linearity and where the curve of turbidimetry at 405 nm starts to increase. [2] ThT Fluorescence in percentage respect to the control sample, after 24 h of aggregation at 37 °C, the concentration was 5 μM for β-amyloids and 25 μM for polyphenols in PBS + 1% DMSO.

β-Amyloid aggregation is known to start when a change of the secondary structure from α-helix (in the membrane environment) or coil (in basic environment) to β-sheet conformation takes place [35]. In the presence of thioflavin-T, β-sheet formation results in a strong increase of fluorescence of the solution. Ideally, inhibitors of aggregation should decrease the maximum plateau of fluorescence obtained in the control experiment. Table 1 shows the relative decrease (or increase) of plateau achieved with compounds **8a,h**, **9**, **13a,b**, **17a,b**.

From the results collected we can draw some useful information. First, as long as we consider Aβ1-42, the lead compound **8a** remains the best one. In particular, substitution of the *t*-Bu group of the isocyanide with a hydroxybenzyl group is highly deleterious, leading to a complete loss of the inhibitory effect (see entries 8 and 9). Only substitution with a hydroxyphenyl group is accepted (entry 6) although a decrease of activity is observed compared to the lead compound **8a**. Also replacing of *p*-hydroxybenzaldehyde with an aliphatic counterpart is detrimental (entry 7).

For AβpE3-42, as already experienced in our previous work, even small differences seem to have a significant effect. The best compound, among those prepared in this second campaign, is **8c**, which turned out to be slightly superior to our lead **8b**. Moreover, this molecule is one of the few, together with **8b**, that shows acceptable activity towards both proteins. Therefore, we think that it is worth of further investigation through in vivo assays. The difference, compared to **8b**, is the amine component, which is a hydroxybenzyl instead of a hydroxyphenyl group. Also, for AβpE3-42, the compounds derived from *iso*-butyraldehyde did not inhibit aggregation. On the contrary, they seem, especially **8f**, to even favor β-sheet formation.

Among the β-lactams, only the one containing the *p*-hydroxyphenyl group showed some activity (entry 14). It was equally active on both proteins and may be a good starting point for further refinement.

It was interesting to see if the different configuration of the new stereogenic center created during the Ugi MCR could influence activity. The results achieved with diastereomers **13a** and **13b** shows a negligible effect on Aβ1-42, but a significant difference in behavior for AβpE3-42. It is worth noting that **13b**, which has just an additional methyl, displays a higher activity for the truncated peptide than **8a**, but we do not know whether this is due to its enantiomeric purity or for the presence of the methyl group.

Finally, it should be noted that compound **9**, lacking any phenolic group, had no inhibitory activity at all. On the contrary, it seems to favor aggregation of Aβ1-42. Furthermore, compounds **8f** and **17a**, that are just monophenols, behaved poorly, stressing the need for a polyphenolic system. The best candidates found so far (**8a**, **8b**, **8c**) are indeed either diphenols or triphenols.

3. Materials and Methods

3.1. General Information

NMR spectra (see Supplementary Materials) were recorded on a Gemini 300 Mhz instrument (Varian, Palo Alto, CA, USA) at r.t. in CDCl$_3$ or in DMSO-d_6 at 300 MHz (^{1}H), and 75 MHz (^{13}C), using, as internal standard, TMS (^{1}H NMR in CDCl$_3$; 0.000 ppm) or the central peak of DMSO (^{1}H-NMR: 2.506 ppm; ^{13}C-NMR: 39.43 ppm) or the central peak of CDCl$_3$ (^{13}C in CDCl$_3$; 77.02 ppm). Chemical shifts are reported in ppm (δ scale). Peak assignments were made with the aid of gCOSY and gHSQC experiments. In ABX system, the proton A is considered upfield and B downfield. IR spectra were recorded as solid, oil, or foamy samples, with the ATR (attenuated total reflectance) technique. TLC analyses were carried out on silica gel plates and viewed at UV (λ = 254 nm or 360 nm) and developed with Hanessian stain (dipping into a solution of (NH$_4$)$_4$MoO$_4$·4H$_2$O (21 g) and Ce(SO$_4$)$_2$·4H$_2$O (1 g) in H$_2$SO$_4$ (31 mL) and H$_2$O (469 mL) and warming). R_f values were measured after an elution of 7–9 cm. HRMS: samples were analyzed with a Synapt G2 QToF mass spectrometer (Waters, Milford, MA, USA). MS signals were acquired from 50 to 1200 *m/z* in either ESI positive or negative ionization mode. Column chromatography was done with the "flash" methodology by using 220–400 mesh silica. Petroleum ether (40–60 °C) is abbreviated as PE. All reactions employing dry solvents were carried out under nitrogen. After extractions, the aqueous phases were always re-extracted 2 times with the appropriate organic solvent, and the organic extracts were always dried over Na$_2$SO$_4$ and filtered before evaporation to dryness.

Due to a tendency to partially degrade, the free phenols (**8c–h**, **13a,b**, **18a,b**) were fully characterized and stored in the acetylated form (**7c–h**, **12a,b**, **17a,b**) and then deprotected shortly before use through procedure B, checking the purity by ^{1}H-NMR and HPLC. HPLC analyses were carried out on a HP-1100 system (Agilent, Santa Clara, CA, USA) equipped with a Phenyl C6 reverse phase column (150 × 3 mm, 3 μm) at 25 °C with flow = 0.34 mL/min. Gradient from CH$_3$CN/H$_2$O 40:60 (time 0) to pure CH$_3$CN (time 10). detection was done with a DAD detector at 330 nm. Compounds **2**, **3** and **5** and 4-(allyloxy)phenyl isocyanide were prepared as previously described [25]. Known [36] 2-(allyloxy)benzylamine was prepared as previously described [32].

3.2. Syntheses

(R, S)-(E)-3-(4-(Allyloxy)-3-methoxyphenyl)-N-(3-(allyloxy)benzyl)-N-(1-(4-(allyloxy)phenyl)-2-(tert-butylamino)-2-oxoethyl)acrylamide (**6c**). A solution of aldehyde **3** (200 mg, 1.23 mmol) in dry trifluoroethanol (TFE, 2.5 mL) and dry ethanol (2.5 mL) was treated with amine **2** (221 mg, 1.35 mmol) and freshly activated powdered 3 Å molecular sieves (62 mg). The suspension was stirred for 8 h at rt. Then, allylated ferulic acid **5** (318 mg, 1.36 mmol) and *tert*-butyl isocyanide (154 μL, 1.36 mmol) were added. After stirring for 48 h at rt, the mixture was diluted with CH$_2$Cl$_2$/MeOH 1:1 and filtered through a Celite cake. After evaporation of the solvent, the crude was taken up in EtOAc and washed

with saturated aqueous NaHCO$_3$ to remove excess of 5, and then with brine. Evaporation of the organic phase to dryness, followed by chromatography (PE:EtOAc 7:3) gave pure **6c** as a white foam (667 mg, 87%). Unreacted aldehyde **3** (24 mg) was also recovered. Yield based on unrecovered starting material: 98%. R$_f$ 0.61 (PE/EtOAc 60:40). ^{1}H-NMR (CDCl$_3$): δ 7.65 (d, *J* = 15.1 Hz, 1 H, C*H*=CHCO), 7.33 (d, *J* = 8.2 Hz, 2 H), 7.19 (broad s, 1 H), 7.08 (t, *J* = 7.8 Hz, 1 H), 6.91 (d, *J* = 7.8 Hz, 1 H), 6.80 (d, *J* = 8.1 Hz, 1 H), 6.78 (d, *J* = 8.2 Hz, 2 H), 6.67 (d, *J* = 8.2 Hz, 1 H), 6.65 (d, *J* = 8.2 Hz, 1 H), 6.54 (d, *J* = 15.1 Hz, 1 H, CH=C*H*CO), 6.52 (s, 1 H), 6.13 (s, 1 H, C*H*N), 6.11–5.89 (m, 3 H, C*H*=CH$_2$), 5.63 (s, 1 H, N*H*), 5.44–5.19 (m, 6 H, CH=C*H*$_2$), 4.85 (d, *J* = 17.9 Hz, 1 H, C*H*HAr), 4.64 (d, *J* = 17.9 Hz, 1 H, CH*H*Ar), 4.60 (d, *J* = 5.4 Hz, 2 H, C*H*$_2$CH=CH$_2$), 4.48 (d, *J* = 4.9 Hz, 2 H, C*H*$_2$CH=CH$_2$), 4.36 (d, *J* = 4.2 Hz, 2 H, C*H*$_2$CH=CH$_2$), 3.78 (s, 3 H, OC*H*$_3$), 1.35 (s, 9 H, C(C*H*$_3$)$_3$). ^{13}C-NMR (CDCl$_3$): δ 169.3, 168.3 (C=O), 158.7, 158.6, 149.5, 149.3, 140.5, 128.4, 127.6 (quat.), 143.2 (CH=CHCO), 133.1, 133.0, 132.8 (CH=CH$_2$), 131.1 (×2), 129.3, 121.9, 118.7, 114.8 (×2), 113.4, 112.8, 112.4, 110.0 (aromatic CH), 118.2, 117.8, 117.6 (CH=CH$_2$), 116.5 (CH=CHCO), 69.7, 68.7, 68.6 (CH$_2$CH=CH$_2$), 61.9 (CHN), 55.8 (OCH$_3$), 51.6 (C(CH$_3$)$_3$), 49.5 ((CH$_2$Ar), 28.7 (C(CH$_3$)$_3$). HRMS: *m/z* (ESI+): 625.3270 (M + H$^+$). C$_{38}$H$_{45}$N$_2$O$_6$ requires 625.3278.

(R,S)-(E)-3-(4-(Acetoxy)-3-methoxyphenyl)-N-(3-(acetoxy)benzyl)-N-(1-(4-(acetoxy)phenyl)-2-(tert-butylamino)-2-oxoethyl)acrylamide (**7c**). A solution of triallyl derivative **6c** (252 mg, 404 μmol), Pd(PPh$_3$)$_2$Cl$_2$ (13 mg, 18.3 μmol, 0.015 eq. calculated on the number of allyl groups), and ammonium formate (114 mg, 1.81 mmol, 1.5 eq. calculated on the number of allyl groups) in dry CH$_3$CN (3.6 mL) in a pressure tube was first flushed with argon and then sealed and heated at 81 °C for 24 h. A tlc showed that the reaction was complete. The resulting dark solution was diluted with EtOAc and saturated aqueous NaHCO$_3$. The phases were separated and the aqueous one re-extracted three times with EtOAc. The organic phases were washed with brine and evaporated to dryness. The crude was taken up in pyridine (1.45 mL, 18.0 mmol) and treated with acetic anhydride (1.45 mL, 15.3 mmol) and stirred at rt for 3 h. The solution was diluted with EtOAc (20 mL), water (20 mL), and 2 M aqueous HCl (10 mL), checking that the resulting pH is <2 (otherwise more HCl is added). The phases were separated, and the aqueous one re-extracted three times with EtOAc. The organic extracts were washed with brine, evaporated to dryness and chromatographed (PE/EtOAc 60:40 + 2% EtOH) to give pure **7c** as a white powder (221 mg, 87%). R$_f$ 0.30 (PE/EtOAc 60:40). IR: υ$_{max}$ 3318, 2969, 2936, 1761, 1680, 1647, 1601, 1538, 1505, 1453, 1417, 1365, 1300, 1256, 1189, 1155, 1121, 1079, 1031, 1012, 977, 946, 908, 869, 829, 795, 750, 731, 697, 643 cm^{-1}. ^{1}H-NMR (CDCl$_3$): δ 7.72 (d, *J* = 15.3 Hz, 1 H, C*H*=CHCO), 7.38 (d, *J* = 8.2 Hz, 2 H), 7.17 (t, *J* = 7.8 Hz, 1 H), 7.10–6.79 (m, 7 H), 6.75 (s, 1 H), 6.61 (d, *J* = 15.3 Hz, 1 H, CH=C*H*CO), 6.10 (s, 1 H, C*H*N), 5.70 (s, 1 H, N*H*), 4.91 (d, *J* = 18.0 Hz, 1 H, C*H*HAr), 4.65 (d, *J* = 18.0 Hz, 1 H, CH*H*Ar), 3.76 (s, 3 H, OC*H*$_3$), 2.29, 2.27, 2.25 (3 s, 3 × 3 H, C*H*$_3$CO), 1.36 (s, 9 H, C(C*H*$_3$)$_3$). ^{13}C-NMR (CDCl$_3$): δ 169.1, 168.8, 168.7, 168.0 (C=O), 151.2, 150.8, 150.7, 141.0, 140.0, 134.0, 132.5 (quat.), 143.5 (CH=CHCO), 130.9 (×2), 129.4, 123.5, 123.1, 121.8 (×2), 120.9, 120.2, 119.5, 111.4 (aromatic CH), 118.2 (CH=CHCO), 61.8 (CHN), 55.8 (OCH$_3$), 51.8 (C(CH$_3$)$_3$), 49.2 (CH$_2$Ar), 28.6 (C(CH$_3$)$_3$), 21.1 (×2), 20.6 (CH$_3$CO). HRMS: *m/z* (ESI+): 631.2664 (M + H$^+$). C$_{35}$H$_{39}$N$_2$O$_9$ requires 631.2656.

(R, S)-(E)-N-(2-(tert-Butylamino)-1-(4-(hydroxy)phenyl)-2-oxoethyl)-N-(3-hydroxybenzyl)-3-(4-hydroxy-3-methoxyphenyl)acrylamide (**8c**). A 0.2 M solution of MeONa in MeOH is freshly prepared treating MeOH with solid Na (4.6 g per liter of MeOH) under nitrogen. Triacetate **7c** (100 mg, 160 μmol) is treated with such solution (4.8 mL, 920 μmol, 2 eq. for each acetyl group). After stirring for 2 h at rt, the solution is treated with dry Amberlyst® 15 (4.7 mmol/g) (freshly thoroughly washed with dry methanol) so that the resulting pH is around 4–5 (about 200 mg of dry resin). The resin was filtered off, washing with methanol and the resulting filtrate evaporated to dryness, to afford **8c**, pure enough for biochemical assays (80 mg, quantitative). The purity by HPLC (for conditions see the general remarks) was 99%. R$_f$ 0.18 (PE/EtOAc 50:50) ^{1}H-NMR (DMSO-*d*$_6$, 90 °C): δ 9.07, 8.98, 8.84 (3 broad s, 3 × 1 H, O*H*), 7.39 (d, *J* = 15.3 Hz, 1 H, C*H*=CHCO), 7.33 (s, 1 H, N*H*), 7.11 (d, *J* = 8.4 Hz, 2 H), 6.98 (s, 1 H), 6.94 (d, *J* = 7.5 Hz, 1 H), 6.93 (broad t, 1 H), 6.75 (d, *J* = 8.1 Hz, 1 H), 6.73–6.68 (broad t, 1 H), 6.68

(d, J = 8.4 Hz, 2 H), 6.58–6.46 (m, 3 H, CH=CHCO and 2 aromatic CH), 5.95 (s, 1 H, CHN), 4.76 (d, J = 16.8 Hz, 1 H, CHHAr), 4.47 (d, J = 16.8 Hz, 1 H, CHHAr), 3.77 (s, 3 H, OCH_3), 1.25 (s, 9 H, C(CH_3)$_3$).

(R, S)-(E)-3-(4-(Allyloxy)-3-methoxyphenyl)-N-(2-(allyloxy)benzyl)-N-(1-(4-(allyloxy)phenyl)-2-(tert-butylamino)-2-oxoethyl)acrylamide (**6d**). This compound was prepared starting from 194 mg of aldehyde **2**, following the same procedure above described for **6c**. Pure **6d** was obtained after chromatography with PE/EtOAc 70:30 + 2% EtOH. Yield: 471 mg (63%). Also 59 mg of unreacted aldehyde was recovered from the column (yield from unrecovered starting aldehyde = 91%). Slightly yellow foam. R_f 0.36 (PE/EtOAc 70:30). ^{1}H-NMR (CDCl$_3$): δ 7.67 (d, J = 15.3 Hz, 1 H, CH=CHCO), 7.29 (d, J = 8.2 Hz, 2 H), 7.19 (broad s, 1 H), 7.09 (t, J = 7.0 Hz, 1 H), 6.93 (d, J = 8.3 Hz, 1 H), 6.85–6.67 (m, 6 H), 6.56 (d, J = 15.3 Hz, 1 H, CH=CHCO), 6.52 (s, 1 H), 6.12–5.92 (m, 3 H, CH=CH$_2$), 5.86 (s, 1 H, CHN), 5.73 (s, 1 H, NH), 5.43–5.23 (m, 6 H, CH=CH_2), 4.86 (d, J = 18.4 Hz, 1 H, CHHAr), 4.69 (d, J = 18.4 Hz, 1 H, CHHAr), 4.60 (dt, J = 5.4 (d), 1.2 (t) Hz, 2 H, CH_2CH=CH$_2$), 4.55–4.40 (m, 4 H, CH_2CH=CH$_2$), 3.78 (s, 3 H, OCH_3), 1.33 (s, 9 H, C(CH_3)$_3$). ^{13}C-NMR (CDCl$_3$): δ 169.1, 168.3 (C=O), 158.4, 155.0, 149.4, 149.3, 128.6, 127.5, 120.6 (quat.), 143.1 (CH=CHCO), 133.2, 133.0, 132.9 (CH=CH$_2$), 131.0 (×2), 127.9, 127.8, 121.6, 120.6, 114.5 (×2), 112.8, 110.9, 110.6 (aromatic CH), 118.2, 117.7, 117.4 (CH=CH$_2$), 116.5 (CH=CHCO), 69.7, 68.7 (×2) (CH$_2$CH=CH$_2$), 63.1 (CHN), 55.9 (OCH$_3$), 51.4 (C(CH$_3$)$_3$), 44.9 ((CH$_2$Ar), 28.7 (C(CH$_3$)$_3$). HRMS: *m/z* (ESI+): 625.3270 (M + H$^+$). C$_{38}$H$_{45}$N$_2$O$_6$ requires 625.3263.

(R, S)-(E)-3-(4-(Acetoxy)-3-methoxyphenyl)-N-(2-(acetoxy)benzyl)-N-(1-(4-(acetoxy)phenyl)-2-(tert-butylamino)-2-oxoethyl)acrylamide (**7d**). It was prepared from 438 mg of **6d** following the same procedure employed for **7c**. Pure **7d** was obtained after chromatography (PE/EtOAc 60:40 + 1% EtOH) (267 mg, 60%). White foam. R_f 0.27 (PE/EtOAc 60:40). IR: υ$_{max}$ 3322, 2967, 2930, 1759, 1681, 1649, 1602, 1542, 1506, 1454, 1418, 1367, 1302, 1259, 1190, 1156, 1122, 1093, 1032, 1011, 980, 951, 907, 826, 751, 664, 645 cm^{-1}. ^{1}H-NMR (CDCl$_3$): δ 7.68 (d, J = 15.3 Hz, 1 H, CH=CHCO), 7.45 (d, J = 8.3 Hz 2 H), 7.16 (broad t, 1 H), 7.10–6.85 (m, 8 H), 6.53 (d, J = 15.3 Hz, 1 H, CH=CHCO), 6.15 (s, 1 H, CHN), 5.63 (s, 1 H, NH), 4.79 (d, J = 18.4 Hz, 1 H, CHHAr), 4.56 (d, J = 18.4 Hz, 1 H, CHHAr), 3.74 (s, 3 H, OCH_3), 2.30, 2.28, 2.27 (3 s, 3 × 3 H, CH_3CO), 1.35 (s, 9 H, C(CH_3)$_3$). ^{13}C-NMR (CDCl$_3$): δ 169.22, 169.16, 168.8, 168.5, 168.1 (C=O), 151.2, 150.7, 150.7, 147.4, 141.0, 134.0, 132.7 (quat.), 143.5 (CH=CHCO), 130.8 (×2), 130.2, 128.1, 127.5, 126.2, 122.9, 121.9 (×2), 121.8, 111.0 (aromatic CH), 117.7 (CH=CHCO), 61.8 (CHN), 55.8 (OCH$_3$), 51.8 (C(CH$_3$)$_3$), 44.4 (CH$_2$Ar), 28.6 (C(CH$_3$)$_3$), 21.1 (×2), 20.6 (CH$_3$CO). HRMS: *m/z* (ESI+): 631.2678 (M + H$^+$). C$_{35}$H$_{39}$N$_2$O$_9$ requires 631.2656.

(R,S)-(E)-N-(2-(tert-Butylamino)-1-(4-(hydroxy)phenyl)-2-oxoethyl)-3-(4-hydroxy-3-methoxyphenyl)-N-(2-(hydroxy)benzyl)acrylamide (**8d**). It was prepared from triacetate **7c** (100 mg, 160 μmol) following the same procedure used for **8c**. Yield: 74.7 mg (93%). The purity by HPLC (for conditions see the general remarks) was 92%. R_f 0.18 (PE/EtOAc 50:50) ^{1}H-NMR (DMSO-d_6, 90 °C): δ 9.33 (broad s, 1 H, OH), 9.08 (broad s, 2 H, OH), 7.40 (d, J = 15.3 Hz, 1 H, CH=CHCO), 7.13 (d, J = 8.5 Hz, 2 H), 7.03–6.58 (m, 9 H), 5.95 (s, 1 H, CHN), 4.75 (d, J = 17.0 Hz, 1 H, CHHAr), 4.50 (d, J = 17.0 Hz, 1 H, CHHAr), 3.77 (s, 3 H, OCH_3), 1.26 (s, 9 H, C(CH_3)$_3$).

(R,S)-(E)-3-(4-(Acetoxy)-3-methoxyphenyl)-N-(1-(4-(acetoxy)phenyl)-2-((4-(acetoxy)phenyl)amino)-2-oxoethyl)-N-benzylacrylamide (**7e**). Triallyl derivative **6e** was prepared from 97 mg of aldehyde **3** (600 μmol), 105 mg of 4-(allyloxy)phenyl isocyanide [25] (660 μmol), 154.5 mg of acid **5** (660 μmol) and 72 μL of benzylamine (660 μmol) following the same procedure employed for **7c**. However, after 48 h, both aldehyde **3** and the starting isocyanide were visible at TLC. The reaction was worked out anyway. Chromatography (PE/EtOAc 60:40 + 1 % EtOH) gave pure **6e** (127 mg, 33%). Starting aldehyde (43 mg) was also recovered. Yield based on non-recovered starting aldehyde = 59%. This triallyl derivative **6e**, pure at TLC, was not fully characterized, but directly converted into **7e**, following the same procedure described for **7c**. Chromatography (PE/EtOAc 70:30 + 3% EtOH) gave pure **7e** (94 mg, 73%). White foam. R_f 0.50 (PE/EtOAc 50:50). IR: υ$_{max}$ 3282, 3070, 2988, 1756, 1697, 1621, 1595, 1546, 1505, 1494, 1453, 1409, 1367, 1310, 1187, 1163, 1106, 1075, 1046, 1014, 966, 909, 846, 757, 737, 700, 675, 634, 611 cm^{-1}.

^{1}H-NMR (CDCl$_3$): δ 8.52 (s, 1 H, N*H*), 7.66 (d, *J* = 15.2 Hz, 1 H, C*H*=CHCO), 7.50 (d, *J* = 8.7 Hz 2 H), 7.45 (d, *J* = 8.4 Hz 2 H), 7.25–7.14 (m, 3 H), 7.12–6.89 (m, 8 H), 6.82 (s, 1 H), 6.68 (d, *J* = 15.3 Hz, 1 H, CH=C*H*CO), 6.30 (s, 1 H, C*H*N), 4.93 (d, *J* = 17.7 Hz, 1 H, C*H*HPh), 4.73 (d, *J* = 17.7 Hz, 1 H, CH*H*Ph), 3.74 (s, 3 H, OC*H*$_3$), 2.29, 2.27 (×2) (3 s, 3 x 3 H, C*H*$_3$CO). ^{13}C-NMR (CDCl$_3$): δ 169.5, 169.1, 168.8, 168.4, 168.0 (C=O), 151.2, 150.8, 146.9, 141.0, 137.7, 135.5, 133.8, 131.8 (quat.), 143.7 (CH=CHCO), 130.8 (×2), 128.6 (×2), 127.3, 126.2 (×2), 123.1, 121.9 (×2), 121.8 (×2), 120.9 (×2), 120.8, 111.5 (aromatic CH), 118.1 (CH=CHCO), 63.1 (CHN), 55.8 (OCH$_3$), 50.1 (CH$_2$Ph), 21.1 (×2), 20.6 (CH$_3$CO). HRMS: *m/z* (ESI+): 651.2371 (M + H$^+$). C$_{37}$H$_{35}$N$_2$O$_9$ requires 631.2343.

(R,S)-(E)-N-Benzyl-3-(4-hydroxy-3-methoxyphenyl)-N-(1-(4-hydroxyphenyl)-2-((4-hydroxyphenyl)-amino)-2-oxoethyl)acrylamide (**8e**). It was prepared from triacetate **7e** (110 mg, 169 µmol) following the same procedure used for **8c**, Yield: 74.2 mg (84%). The purity by HPLC (for conditions see the general remarks) was 96%. R$_f$ 0.17 (PE/EtOAc 50:50) ^{1}H-NMR (DMSO-*d*$_6$, 90 °C): δ 9.67 (s, 1 H, O*H*), 9.14 (s, 1 H, O*H*), 8.99 (broad s, 1 H, N*H*), 8.84 (s, 1 H, O*H*), 7.42 (d, *J* = 15.3 Hz, 1 H, C*H*=CHCO), 7.32 (d, *J* = 9.0 Hz, 2 H), 7.22–7.05 (m, 7 H), 6.96 (broad s, 1 H), 6.90 (broad d, *J* = 8.1 Hz, 1 H), 6.78–6.65 (m, 6 H), 6.20 (s, 1 H, C*H*N), 4.89 (d, *J* = 17.1 Hz, 1 H, C*H*HAr), 4.63 (d, *J* = 17.1 Hz, 1 H, CH*H*Ar), 3.75 (s, 3 H, OC*H*$_3$).

(R,S)-(E)-3-(4-(Acetoxy)-3-methoxyphenyl)-N-benzyl-N-(2-(tert-butylamino)-1-isopropyl-2-oxoethyl)acrylamide (**7f**). Allyl derivative **6f** was prepared from 223.2 mg of acid **5** (1.05 mmol) and 1.1 eq. of benzylamine, isobutyraldehyde and *tert*-butyl isocyanide, following the same procedure employed for **7c**. However, after 48 h, acid **5** was still visible at TLC. The reaction was worked out anyway. Chromatography (PE/EtOAc 60:40 + 1% EtOH) gave pure **6f** (155 mg, 34%) This allyl derivative **6f**, pure at TLC, was not fully characterized, but directly converted into **7f**, following the same procedure described for **7c**. Chromatography (PE/EtOAc 70:30) gave pure **7f** (103.8 mg, 67%). White foam. R$_f$ 0.29 (PE/AcOEt 70:30). ^{1}H-NMR (CDCl$_3$): δ 7.63 (d, *J* = 15.3 Hz, 1 H, C*H*=CHCO), 7.35–7.18 (m, 5 H), 6.99–6.88 (m, 2 H), 6.79 (s, 1 H), 6.61 (d, *J* = 15.3 Hz, 1 H, CH=C*H*CO), 6.27 (broad s, 1 H, N*H*), 4.93 (d, *J* = 17.2 Hz, 1 H, C*H*HPh), 4.74 (d, *J* = 17.2 Hz, 1 H, CH*H*Ph), 4.56 (broad, *J* = 6.9 Hz, 1 H, C*H*N), 3.74 (s, 3 H, OC*H*$_3$), 2.51–2.34 (m, 1 H, C*H*(CH$_3$)$_2$), 2.29 (s, 3 H, C*H*$_3$CO), 1.29 (s, 9 H, C(C*H*$_3$)$_3$), 0.99 (d, *J* = 6.4 Hz, 3 H, C*H*$_3$CH), 0.87 (d, *J* = 6.5 Hz, 3 H, C*H*$_3$CH). ^{13}C-NMR (CDCl$_3$): δ 169.2, 168.8, 168.2 (C=O), 151.2, 140.9, 138.3, 134.1 (quat.), 142.5 (CH=CHCO), 128.6 (×2), 127.2, 126.4 (×2), 123.1, 120.8, 111.2 (aromatic CH), 118.8 (CH=CHCO), 65.6 (very broad) CHN, 55.8 (OCH$_3$), 51.4 (C(CH$_3$)$_3$), 48.5 (very broad) (CH$_2$Ph), 28.6 (C(CH$_3$)$_3$), 27.5 (CH(CH$_3$)$_2$), 20.6 (CH$_3$CO), 19.7, 19.0 (CH$_3$CH). HRMS: *m/z* (ESI+): 481.2708 (M + H$^+$). C$_{28}$H$_{37}$N$_2$O$_5$ requires 481.2702.

(R,S)-(E)-N-benzyl-N-(2-(tert-butylamino)-1-isopropyl-2-oxoethyl)-3-(4-hydroxy-3-methoxyphenyl)-acrylamide (**8f**). Prepared from acetate **7f** (89.6 mg, 186 µmol) following the same procedure used for **8c**, Yield: 59.1 mg (73%). The purity by HPLC (for conditions see the general remarks) was 96%. R$_f$ 0.32 (PE/EtOAc 60:40). ^{1}H-NMR (DMSO-*d*$_6$, 90 °C; some signals were still rather broad at this temperature): δ 8.96 (broad s, 1 H, O*H*), 7.41 (d, *J* = 15.3 Hz, 1 H, C*H*=CHCO), 7.41 (broad s, 1 H, N*H*), 7.32–7.24 (m, 5 H), 7.23–7.11 (m, 1 H), 7.10–6.87 (broad m, 2 H), 6.77 (d, *J* = 8.1 Hz, 1 H), 4.94 (broad d, *J* = 16.2 Hz, 1 H, C*H*HPh), 4.75 (d, *J* = 16.2 Hz, 1 H, CH*H*Ph), 3.79 (s, 3 H, OC*H*$_3$), 2.37–2.23 (m, 1 H, C*H*(CH$_3$)$_2$), 1.22 (s, 9 H, C(C*H*$_3$)$_3$), 0.95 (d, *J* = 6.5 Hz, 3 H, C*H*$_3$CH), 0.82 (d, *J* = 6.6 Hz, 3 H, C*H*$_3$CH).

(R,S)-(E)-N-(2-(4-(Acetoxy)benzyl)-1-isopropyl-2-oxoethyl)-3-(4-(Acetoxy)-3-methoxyphenyl)-N-benzylacrylamide (**7g**). Diallyl derivative **6g** was prepared from 357 mg of acid **5** (1.52 mmol), 1 eq. each of benzylamine and isobutyraldehyde and 1.3 eq. of 4-(allyloxy)benzyl isocyanide [32], following the same procedure employed for **7c**. The reaction was worked out as usual after 48 h. As in our previous paper [32], the isocyanide was not stripped at high vacuum due to its volatility/lability and thus the actual amount used was slightly lower. Chromatography (PE/EtOAc 60:40) gave pure **6g** (382 mg, 44%) This diallyl derivative **6g**, pure at TLC, was not fully characterized, but directly converted into **7g**, following the same procedure described for **7c**. Chromatography (PE/EtOAc 70:30) gave pure **7g** (253 mg, 66%).

White foam. R_f 0.37 (PE/EtOAc 60:40). IR: υ_{max} 3290, 3065, 2965, 2874, 1759, 1673, 1647, 1600, 1507, 1453, 1418, 1367, 1301, 1257, 1213, 1190, 1156, 1121, 1080, 1031, 1013, 972, 941, 908, 830, 731, 697, 645 cm^{-1}. ^{1}H-NMR (CDCl$_3$): δ 7.62 (d, *J* = 15.3 Hz, 1 H, C*H*=CHCO), 7.31–7.16 (m, 8 H), 7.02 (d, *J* = 8.7 Hz, 2 H), 6.99-6.90 (m, 2 H), 6.81 (d, *J* = 1.2 Hz, 1 H), 6.63 (d, *J* = 15.3 Hz, 1 H, CH=C*H*CO), 4.83, 4.77 (AB syst., *J* = 17.4 Hz, 2 H, C*H*$_2$N), 4.62 (broad d, *J* = 10.5 Hz, 1 H, C*H*N), 4.43 (dd, *J* = 14.9, 6.1 Hz, 1 H C*H*HNH), 4.27 (dd, *J* = 14.9, 5.7 Hz, 1 H CH*H*NH), 3.76 (s, 3 H, OC*H*$_3$), 2.59–2.43 (m, 1 H, C*H*(CH$_3$)$_2$), 2.30 (s, 3 H, C*H*$_3$CO), 2.28 (s, 3 H, C*H*$_3$CO), 1.00 (d, *J* = 6.4 Hz, 3 H, C*H*$_3$CH), 0.87 (d, *J* = 6.6 Hz, 3 H, C*H*$_3$CH). ^{13}C-NMR (CDCl$_3$): δ 170.1, 169.4, 168.8, 168.5 (C=O), 151.2, 149.8, 141.0, 137.7, 135.8, 133.9 (quat.), 143.0 (*C*H=CHCO), 128.8 (×2), 128.7 (×2), 127.4, 126.4 (×2), 123.1, 121.7 (×2), 120.8, 111.3 (aromatic CH), 118.4 (CH=*C*HCO), 65.8 (very broad) (*C*HN), 55.8 (O*C*H$_3$), 49.2 (very broad) (*C*H$_2$Ph), 42.7 (*C*H$_2$NH), 27.1 (*C*H(CH$_3$)$_2$, 21.1 (*C*H$_3$CO), 20.6 (*C*H$_3$CO), 19.9, 19.2 (*C*H$_3$CH). HRMS: *m/z* (ESI+): 573.2608 (M + H$^+$). C$_{33}$H$_{37}$N$_2$O$_7$ requires 573.2601.

(R,S)-(E)-N-benzyl-N-(2-(4-hydroxybenzyl)-1-isopropyl-2-oxoethyl)-3-(4-hydroxy-3-methoxyphenyl)-acrylamide (**8g**). Prepared from acetate **7g** (104.2 mg, 182 μmol) following the same procedure used for **8c**, Yield: 71.9 mg (73%). The purity by HPLC (for conditions see the general remarks) was 96%. R_f 0.35 (PE/EtOAc 50:50). ^{1}H-NMR (DMSO-d_6, 90 °C; some signals were still rather broad at this temperature): δ 9.00 (broad s, 1 H, O*H*), 8.93 (s, 1 H, N*H*), 8.27 (s, 1 H, O*H*), 7.40 (d, *J* = 15.2 Hz, 1 H, C*H*=CHCO), 7.36–7.14 (m, 7 H), 7.05 (d, *J* = 8.4 Hz, 2 H), 7.05–6.93 (broad m, 2 H), 6.77 (d, *J* = 7.8 Hz, 1 H), 6.70 (d, *J* = 8.4 Hz, 2 H), 4.95–4.55 (m, 3 H, C*H*$_2$Ph and C*H*N), 4.12 (d, *J* = 6.0 Hz, 2 H, C*H*$_2$NH), 3.78 (s, 3 H, OC*H*$_3$), 2.41–2.22 (m, 1 H, C*H*(CH$_3$)$_2$), 0.92 (d, *J* = 6.6 Hz, 3 H, C*H*$_3$CH), 0.81 (d, *J* = 6.6 Hz, 3 H, C*H*$_3$CH).

N-3-(Allyloxy)benzyl formamide. Amine **2**, prepared as previously described [25] (1.10 g, 6.74 mmol) was dissolved in ethyl formate (27 mL) and heated at reflux for 16 h. Evaporation of the solvent gave the title compound in quantitative yield (1.289 g) as a white solid. M.p.: 39.3–41.5 °C. R_f 0.12 (PE/AcOEt 50:50). IR: υ_{max} 3271, 3044, 2933, 2914, 2888, 2863, 2775, 1642, 1614, 1587, 1539, 1486, 1452, 1423, 1388, 1367, 1347, 1310, 1295, 1261, 1236, 1216, 1156, 1098, 1035, 992, 946, 913, 884, 813, 772, 730, 693, 658, 625 cm^{-1}. ^{1}H NMR (CDCl$_3$) (two conformers A and B in 87:13 ratio, are visible): δ: δ 8.24 (A) (s, 0.87 H, C*H*=O), 8.15 (B) (d, *J* = 11.9 Hz, 0.13 H, C*H*=O), 7.30–7.19 (A + B) (m, 1 H), 6.90–6.77 (A + B) (m, 3 H), 6.04 (A + B) (ddt, *J* = 17.1, 10.5 (d), 5.3 Hz (t), 1 H, C*H*=CH$_2$), 5.99 (A + B) (broad s, N*H*), 5.28 (A + B) (dq, *J* = 10.5 (d), 1.2 Hz (q), 1 H CH=C*H*H), 4.56–4.49 (A + B) (m, 2 H, C*H*$_2$CH=CH$_2$), 4.44 (A) (d, *J* = 5.9 Hz, 1.74 H, C*H*$_2$Ar), 4.36 (B) (d, *J* = 6.5 Hz, 0.26 H, C*H*$_2$Ar). ^{13}C-NMR (CDCl$_3$) only the peaks of major conformer are listed): δ 161.0, 158.9 (C=O), 139.1, 129.8 (quat.), 133.0 (*C*H=CH$_2$), 129.8, 120.1, 114.2, 113.8 (aromatic CH), 117.7 (CH=*C*H$_2$), 68.7 (*C*H$_2$CH=CH$_2$), 42.0 (Ar*C*H$_2$). HRMS: *m/z* (ESI+): 192.1017 (M + H$^+$). C$_{11}$H$_{14}$NO$_2$ requires 192.1025.

3-(Allyloxy)benzyl isocyanide. N-3-(Allyloxy)benzyl formamide (386 mg, 2.02 mmol) was dissolved in dry CH$_2$Cl$_2$ (20 mL), cooled to −30 °C and treated with triethylamine (1.30 mL, 9.29 mmol) and POCl$_3$(282 μL, 3.05 mmol). After stirring for 3 h at the same temperature, the brown solution was poured into saturated aqueous NaHCO$_3$, and extracted three times with CH$_2$Cl$_2$. The organic extracts were washed with brine, evaporated to dryness and chromatographed (PE/EtOAc 95:5) to give the pure title compound as a colorless liquid. Due to its partial volatility, it was evaporated only at 15 mbar and 25 °C. The yield (331 mg, 95%) is therefore slightly overestimated, since it probably contains few amounts of solvent. Ths isocyanide was directly used for the next Ugi reaction to give **6h** (see below).

(R, S)-(E)-N-(2-(3-(Acetoxy)benzyl)-1-(4-acetoxyphenyl)-2-oxoethyl)-3-(4-(Acetoxy)-3-methoxyphenyl)-N-benzylacrylamide (**7h**). Triallyl derivative **6h** was prepared from 243 mg of aldehyde **3** (1.50 mmol), 387 mg of acid **5** (1.65 mmol), 1.1 eq. of benzylamine and and freshly prepared 3-(allyloxy)benzyl isocyanide (330 mg, 1.91 mmol), following the same procedure employed for **7c**. The reaction was worked out as usual after 48 h. Chromatography (PE/EtOAc 60:40) gave pure **6h** (580 mg, 59%). Chromatography gave also mg 90 of recovered starting aldehyde **3**. Thus, the yield from unrecovered

starting aldehyde is 93%. This triallyl derivative **6h**, pure at TLC, was not fully characterized, but directly converted into **7h**, following the same procedure described for **7c**. Chromatography (PE/EtOAc 40:60) gave pure **7h** (560 mg, 96%). White foam. R_f 0.51 (PE/EtOAc 50:50). IR: υ_{max} 3297, 3064, 2938, 1759, 1677, 1648, 1601, 1506, 1451, 1418, 1367, 1299, 1257, 1189, 1155, 1121, 1081, 1014, 976, 956, 907, 829, 793, 731, 696, 635 cm^{-1}. ^{1}H-NMR (CDCl$_3$): δ 7.70 (d, J = 15.3 Hz, 1 H, CH=CHCO), 7.42 (d, J = 8.4 Hz, 2 H), 7.31 (t, J = 7.7 Hz, 1 H), 7.25–6.94 (m, 12 H), 6.84 (s, 1 H), 6.64 (d, J = 15.3 Hz, 1 H, CH=CHCO), 6.33 (broad t, J = 5.7 Hz, 1 H, NH), 6.08 (s, 1 H, CHN), 4.89, 4.65 (AB syst., J = 17.8 Hz, 2 H, CH_2Ph), 4.51, 4.46 (AB part of an ABX syst, J_{AB} = 14.6, J_{AX} = 5.0, J_{BX} = 6.0 Hz, CH_2NH), 3.76 (s, 3 H, OCH_3), 2.29 (s, 3 H, CH_3CO), 2.27 (s, 3 H, CH_3CO), 2.26 (s, 3 H, CH_3CO). ^{13}C-NMR (CDCl$_3$): δ 169.5, 169.4, 169.1, 168.8, 168.0 (C=O), 151.1, 150.8, 141.0, 139.7, 137.8, 134.0, 132.2 (quat.), 143.2 (CH=CHCO), 131.0 (×2), 129.6, 128.6 (×2), 127.1, 126.2 (×2), 125.1, 123.0, 121.9 (×2), 120.84, 120.81, 120.6, 111.3 (aromatic CH), 118.3 (CH=CHCO), 62.3 (CHN), 55.8 (OCH$_3$), 50.0 (CH$_2$Ph), 43.2 (CH$_2$NH), 21.1 (CH$_3$CO), 21.0 (CH$_3$CO), 20.6 (CH$_3$CO). HRMS: m/z (ESI+): 665.2522 (M + H$^+$). C$_{38}$H$_{37}$N$_2$O$_9$ requires 665.2499.

(R,S)-(E)-N-Benzyl-N-(2-(3-(hydroxy)benzyl)-1-(4-hydroxyphenyl)-2-oxoethyl)-3-(4-hydroxy-3-methoxyphenyl) acrylamide (**8h**). Prepared from acetate **7h** (100.0 mg, 150 μmol) following the same procedure used for **8c**. Yield: 76.2 mg (94%). The purity by HPLC (for conditions see the general remarks) was 94%. R_f 0.32 (PE/EtOAc 60:40). ^{1}H-NMR (DMSO-d_6, 90 °C) (some signals were still rather broad at this temperature): δ 9.09 (s, 1 H), 8.90 (s, 2 H), 8.27 (broad s, 1 H), 7.41 (d, J = 15.3 Hz, 1 H, CH=CHCO), 7.25–7.02 (m, 9 H), 6.96 (s, 1 H), 6.90 (d, J = 8.1 Hz, 1 H), 6.75 (d, J = 8.1 Hz, 1 H), 6.72–6.61 (m, 5 H), 6.14 (broad s, CHN), 4.88 (d, J = 17.1 Hz, 1 H, CHHPh), 4.61 (d, J = 17.1 Hz, 1 H, CHHPh), 4.26, 4.21 (AB part of an ABX syst, J_{AB} = 15.2, J_{AX} = 5.9, J_{BX} = 6.2 Hz, CH_2NH), 3.76 (s, 3 H, OCH_3).

N-Benzyl-N-(2-(tert-butylamino)-2-oxo-1-phenylethyl)cinnamamide (**9**). Benzaldehyde (153 μL, 1.50 mmol), benzylamine (180 μL, 1.65 mmol), cinnamic acid (245 mg, 1.65 mmol) and *tert*-butyl isocyanide (185 μL, 1.65 mmol) were reacted as described for the synthesis of **6c**. Chromatography (PE: AcOEt 75:25) gave pure **9** as a white foam (571 mg, 89%). R_f 0.36 (PE/EtOAc 75:25). IR: υ_{max} 3316, 3063, 3030, 2971, 2926, 1650, 1596, 1547, 1496, 1470, 1450, 1411, 1392, 1363, 1351, 1331, 1303, 1284, 1253, 1220, 1201, 1189, 1174, 1078, 1032, 997, 976, 947, 915, 892, 860, 841, 804, 768, 758, 740, 723, 695, 642, 622, 615 cm^{-1}. ^{1}H-NMR (CDCl$_3$): δ 7.77 (d, J = 15.3 Hz, 1 H, CH=CHCO), 7.43–7.08 (m, 14 H), 7.01 (broad d, J = 6.9 Hz, 1 H), 6.72 (d, J = 15.3 Hz, 1 H, PhCH=CH), 6.11 (s, 1 H, CHN), 5.68 (s, 1 H, NH), 4.93, 4.69 (AB syst., J = 17.9 Hz, 2 H, CH_2Ph), 1.35 (s, 9 H, C(CH_3)$_3$). ^{13}C-NMR (CDCl$_3$): δ 168.9, 168.2 (C=O), 138.2, 135.3, 135.1 (quat.), 143.7 (CH=CHCO), 129.7 (×3), 128.7 (×2), 128.6 (×2), 128.3 (×3), 127.9, 126.9 (×2), 126.2 (×2) (aromatic CH), 118.2 (CH=CHCO), 62.9 (CHN), 51.7 (C(CH$_3$)$_3$), 49.7 (CH$_2$Ph), 28.6 (C(CH$_3$)$_3$). HRMS: m/z (ESI+): 427.2378 (M + H$^+$). C$_{28}$H$_{31}$N$_2$O$_2$ requires 427.2386.

(E)-N-(1-(4-Acetoxyphenyl)-2-(tert-butylamino)-2-oxoethyl))-3-(4-acetoxy-3-methoxyphenyl)-N-((S)-1-phenylethyl)acrylamide (**12a**) *and (E)-N-(1-(4-Acetoxyphenyl)-2-(tert-butylamino)-2-oxoethyl))-3-(4-acetoxy-3-methoxyphenyl)-N-((S)-1-phenylethyl)acrylamide* (**12b**). Aldehyde **3** (243 mg, 1.50 mmol), protected ferulic acid **5** (387 mg, 1.65 mmol), (S)-α-methylbenzylamine (212.7 μL, 1.65 mmol) and *tert*-butyl isocyanide (185 μL, 1.65 mmol) were reacted as described for the synthesis of **6c**. Chromatography (PE/EtOAc 70:30) gave: the faster running diastereomer **11a** (182.1 mg) (R_f = 0.27, PE/EtOAc 70:30), the slower running diastereomer **11b** (190.4 mg) (R_f = 0.19, PE/EtOAct 70:30), and some mixed fractions (37.6 mg). A second chromatography of the mixed fractions gave additional **11a** (21.7 mg) and **11b** (15.9 mg). Overall yield: 420.1 mg (44%). Diastereomeric ratio = 50:50. The realtive configuration was not established. Diallyl derivatives **11a** and **11b**, pure at TLC, were not fully characterized, but directly independently converted into **12a** and **12b**, following the same procedure described for **7c**.

12a. Obtained in 45% yield (96.3 mg from 212.1 mg of **11a**) after chromatography with PE/EtOAc 60:40). R_f 0.34 (PE/EtOAc 60:40). IR: υ_{max} 3300, 2965, 2930, 1768, 1689, 1644, 1603, 1547, 1508, 1452, 1434, 1416, 1389, 1368, 1337, 1260, 1193, 1152, 1123, 1034, 1014, 978, 946, 909, 883, 846, 828, 794 cm^{-1}. ^{1}H-NMR (CDCl$_3$) (due to the presence of conformers around the tertiary amide, the signals are rather broad and splitting of some signal is present): δ 7.72 (d, *J* = 15.0 Hz, 1 H, C*H*=CHCO), 7.64–7.25 (m, 7 H), 7.10 (d, *J* = 8.7 Hz, 2 H), 6.97–6.78 (m, 2 H), 6.73–6.53 (m, 1 H), 6.44 (d, *J* = 15.0 Hz, 1 H, CH=C*H*CO), 6.59 (minor conformer) and 5.43 (major conformer) (2 very broad s, 1 H, C*H*CH$_3$), 5.00 (s, 1 H, C*H*N), 4.78 (s, 1 H, N*H*), 3.72 (s, 3 H, OC*H*$_3$), 2.29 (s, 6 H, C*H*$_3$CO), 1.53–1.17 (m, 6 H, C*H*$_3$CH and C(C*H*$_3$)$_3$ (minor conformer)), 0.94 (s, 6 H, C(C*H*$_3$)$_3$, major conformer). ^{13}C-NMR (CDCl$_3$): δ 169.2, 168.8, 168.3, 167.6 (C=O), 151.1, 150.0, 142.7, 141.0, 140.9, 134.0, 133.3 (quat.), 142.7 (CH=CHCO), 129.7 (×2), 129.0 (×2), 128.6, 127.9 (×2), 123.0, 121.7, 121.1 (×2), 111.3 (aromatic CH), 119.9 (CH=CHCO), 63.0 (C(CH$_3$)$_3$), 59.7 (CHN), 55.7 (OCH$_3$), 51.5 (CHPh), 28.3 (C(CH$_3$)$_3$), 21.1, 20.6 (CH$_3$CO), 16.3 (CHCH$_3$). HRMS: *m/z* (ESI+): 587.2774 (M + H$^+$). C$_{34}$H$_{39}$N$_2$O$_7$ requires 587.2757.

12b. Obtained in 67% yield (145.5 mg from 216.5 mg of **11a**) after chromatography with PE/EtOAc 60:40). R$_f$ 0.30 (PE/EtOAc 60:40). IR: υ$_{max}$ 3300, 2970, 2936, 1761, 1685, 1646, 1600, 1545, 1506, 1452, 1431, 1417, 1393, 1366, 1339, 1260, 1191, 1154, 1121, 1031, 1012, 975, 945, 908, 882, 844, 827, 792 cm^{-1}. ^{1}H-NMR (CDCl$_3$) (due to the presence of conformers around the tertiary amide, the signals are rather broad and splitting of some signal is present): δ 7.64 (d, *J* = 15.0 Hz, 1 H, C*H*=CHCO), 7.50–7.12 (m, 6 H), 7.10–6.85 (m, 4 H), 6.82 (d, *J* = 15.0 Hz, 1 H, CH=C*H*CO), 6.52 (s, 1 H, N*H*), 5.38 (C*H*CH$_3$), 5.05 (C*H*N), 3.80 (s, 3 H, OC*H*$_3$), 2.31 (s, 3 H, C*H*$_3$CO), 2.25 (s, 3 H, C*H*$_3$CO), 1.92–1.60 (m, 3 H, C*H*$_3$CH), 1.35 (s, 9 H, C(C*H*$_3$)$_3$). ^{13}C-NMR (CDCl$_3$): δ 169.5, 169.1, 168.8, 167.3 (C=O), 151.2, 150.0, 140.9, 139.9, 134.2, 133.9 (quat.), 142.6 (CH=CHCO), 129.7 (×2), 128.5 (×2), 127.8, 127.3 (×2), 123.1, 121.4 (×2), 120.7 (×2), 111.2 (aromatic CH), 119.3 (CH=CHCO), 64.2 (CHN), 60.4 (C(CH$_3$)$_3$), 55.8 (OCH$_3$ and CHPh), 28.3 (C(CH$_3$)$_3$), 21.1, 20.6 (CH$_3$CO), 16.3 (CHCH$_3$). HRMS: *m/z* (ESI+): 587.2768 (M + H$^+$). C$_{34}$H$_{39}$N$_2$O$_7$ requires 587.2757.

(E)-N-(2-(tert-Butylamino)-1-(4-hydroxyphenyl)-2-oxoethyl))-3-(4-hydroxy-3-methoxyphenyl)-N-((S)-1-phenylethyl)acrylamide **(13a)** and *(E)-N-(2-(tert-Butylamino)-1-(4-hydroxyphenyl)-2-oxoethyl))-3-(4-hydroxy-3-methoxyphenyl)-N-((S)-1-phenylethyl)acrylamide* **(13b)**. They were independently prepared respectively from diacetates **12a** and **12b**, following the same procedure used for **8c**.

13a. Yield: 90%. The purity by HPLC (for conditions see the general remarks) was 94%. R$_f$ 0.29 (PE/EtOAc 50:50). ^{1}H-NMR (DMSO-*d*$_6$, 90 °C): δ 9.09 (broad s, 2 H, O*H*), 7.50 (d, *J* = 7.5 Hz, 2 H), 7.36–7.17 (m, 6 H, aromatic C*H* and C*H*=CHCO), 6.94 (s, 1 H, N*H*), 6.81–6.69 (m, 5 H, aromatic C*H*), 6.50 (d, *J* = 15.3 Hz, CH=C*H*CO), 5.79 (broad m, 1 H, C*H*CH$_3$), 5.45 (s, 1 H, C*H*N), 3.74 (s, 3 H, OC*H*$_3$), 1.49 (d, *J* = 7.0 Hz, 3 H, C*H*$_3$CH), 1.14 (s, 9 H, C(C*H*$_3$)$_3$).

13b. Yield: 85%. The purity by HPLC (for conditions see the general remarks) was 98%. R$_f$ 0.34 (PE/EtOAc 50:50). ^{1}H-NMR (DMSO-*d*$_6$, 90 °C): δ 9.02 (broad s, 2 H, O*H*), 7.65 (s, 1 H, N*H*), 7.41–7.16 (m, 7 H, aromatic C*H* and C*H*=CHCO), 6.86 (d, *J* = 7.9 Hz, 2 H), 6.81–6.67 (m, 3 H, aromatic C*H*), 6.60 (d, *J* = 8.6 Hz, 1 H), 6.51 (d, *J* = 15.3 Hz, CH=C*H*CO), 5.68 (broad m, 1 H, C*H*CH$_3$), 5.35 (s, 1 H, C*H*N), 3.75 (s, 3 H, OC*H*$_3$), 1.66 (d, *J* = 7.0 Hz, 3 H, C*H*$_3$CH), 1.35 (s, 9 H, C(C*H*$_3$)$_3$).

(R,S)-(E)-3-(4-(Allyloxy)-3-methoxyphenyl)-N-benzyl-N-(1-(tert-butyl)-2-oxoazetidin-3-yl)acrylamide **(16a)**. A solution of glycolaldehyde dimer (135 mg, 1.125 mmol) in dry ethanol (7.5 mL), was treated with benzylamine (246 µL, 2.25 mmol) and freshly activated powdered 3 Å molecular sieves (115 mg). The suspension was stirred for 6 h at rt. Then, allylated ferulic acid **5** (350 mg, 1.50 mmol) and *tert*-butyl isocyanide (220 µL, 1.94 mmol) were added. After stirring for 48 h at rt, the mixture was diluted with CH$_2$Cl$_2$/MeOH 1:1 and filtered through a Celite cake. After evaporation of the solvent, the crude was taken up in EtOAc and washed with saturated aqueous NaHCO$_3$ to remove excess of **5**, and then with brine. Evaporation of the organic phase to dryness, followed by chromatography (PE/EtOAc 50:50 + 2% EtOH) gave pure **15a** as a brownish foam (292 mg, 42%). This compound (277 mg, 0.59 mmol) was taken up in dry DMF (2.0 mL) cooled at 0 °C and treated with sulfonyl diimidazole (177 mg, 0.89 mmol) and NaH (60% in mineral oil) (36 mg, 0.89 mmol). After stirring for 2.5 h at rt, the mixture was diluted with saturated aqueous NH$_4$Cl (30 mL) and Et$_2$O/DCM 20:1. The phases were separated, and the aqueous one re-extracted three times with Et$_2$O/DCM 20:1. The organic phases were washed with brine (×4),

evaporated to dryness and chromatographed (PE/EtOAc 3:2 + 2% EtOH) to give pure **16a** as a yellowish oil (115 mg, 43%). R_f 0.49 (PE/EtOAc 50:50 + 2% EtOH). ^{1}H- NMR (CDCl$_3$): δ 7.72 (d, J = 15.3 Hz, 1 H, C*H*=CHCO), 7.41–7.24 (m, 5 H), 7.01 (d, J = 7.8 Hz, 1 H), 6.91 (s, 1 H), 6.82 (d, J = 8.4 Hz, 1 H), 6.66 (d, J = 15.3 Hz, 1 H, CH=C*H*CO), 6.06 (ddt, J = 5.4 (t), 10.6, 17.6 Hz (d), 1 H, C*H*=CH$_2$), 5.40 (dq, J = 17.6 (d), 0.9 Hz (q), 1 H, C*H*H=CH), 5.30 (dq, J = 10.6 (d), 0.9 Hz (q), 1 H, CH*H*=CH), 4.97 (broad s, 1 H, C*H*N), 4.86, 4.78 (AB syst., J = 17.4 Hz, 2 H, C*H*$_2$Ph), 4.62 (d, J = 5.4 Hz, 2 H, C*H*$_2$CH=CH$_2$), 3.85 (s, 3 H, OC*H*$_3$), 3.39 (t, J = 5.1 Hz., C*H*HCO), 3.18 (broad s, 1 H, CH*H*CO), 1.24 (s, 9 H, C(C*H*$_3$)$_3$). ^{13}C-NMR (CDCl$_3$): δ 167.7, 164.8 (C=O), 149.8, 149.4, 137.6, 128.2 (quat.), 144.2 (CH=CHCO), 132.8 (CH=CH$_2$), 128.9 (×2), 127.8, 126.7 (×2), 121.9, 112.9, 110.4 (aromatic CH), 118.3 (CH=CH$_2$), 114.7 (CH=CHCO), 69.7 (CH$_2$CH=CH$_2$), 61.3 (CHN), 56.0 (OCH$_3$), 53.2 (C(CH$_3$)$_3$), 51.0 (CH$_2$Ph), 43.7 ((CH$_2$CO), 27.4 (C(CH$_3$)$_3$). HRMS: m/z (ESI+): 449.2443 (M + H$^+$). C$_{27}$H$_{33}$N$_2$O$_4$ requires 449.2440.

(R,S)-(E)-3-(4-(Allyloxy)-3-methoxyphenyl)-N-(1-(4-allyloxyphenyl)-2-oxoazetidin-3-yl)-N-benzyl-acrylamide (**16b**). Prepared following the same procedure above described for **16a**. Starting from 350 mg of allylated ferulic acid **5** (1.50 mmol), pure **15b** was obtained after chromatography (PE/EtOAc from 50:50 + 1% AcOH to 50:50 + 2% EtOH) as yellow-green oil (222 mg, 28%). Then, this alcohol (210 mg, 0.40 mmol) was converted into **16b** as described for **16a**. Chromatography: PE/EtOAc from 75:25 + 2% EtOH to 50:50 + 4% EtOH. Yield: 139 mg, 66%. R_f 0.35 (PE/EtOAc 70:30 + 2% EtOH). ^{1}H-NMR (CDCl$_3$): δ 7.73 (d, J = 15.2 Hz, 1 H, C*H*=CHCO), 7.40–7.22 (m, 7 H), 7.00 (d, J = 8.3 Hz, 1 H), 6.92–6.78 (m, 4 H), 6.65 (d, J = 15.2 Hz, 1 H, CH=C*H*CO), 6.04 (ddt, J = 5.4 (t), 10.6, 17.6 Hz (d), 1 H, C*H*=CH$_2$), 6.03 (ddt, J = 5.4 (t), 10.6, 17.6 Hz (d), 1 H, C*H*=CH$_2$), 5.39 (dq, J = 17.4 (d), 1.5 Hz (q), 2 H, C*H*H=CH), 5.30 (dq, J = 10.6 (d), 0.9 Hz (q), 1 H, CH*H*=CH), 5.28 (dq, J = 10.6 (d), 0.9 Hz (q), 1 H, CH*H*=CH), 5.25 (broad s, 1 H, C*H*N), 4.83 (broad s, 2 H, C*H*$_2$Ph), 4.62 (dt, J = 5.4 (d), 1.4 Hz (t), 2 H, C*H*$_2$CH=CH$_2$), 4.50 (dt, J = 5.3 (d), 1.5 Hz (t), 2 H, C*H*$_2$CH=CH$_2$), 3.83 (s, 3 H, OC*H*$_3$), 3.79 (t, J = 5.7 Hz., C*H*HCO), 3.63 (broad s, 1 H, CH*H*CO). ^{13}C-NMR (CDCl$_3$): δ 167.7, 162.9 (C=O), 155.2, 149.9, 149.4, 137.1, 131.8, 128.0 (quat.), 144.6 (CH=CHCO), 133.1, 132.7 (CH=CH$_2$), 129.0 (×2), 127.9, 126.7 (×2), 122.1, 117.9 (×2), 115.3 (×2), 112.9, 110.3 (aromatic CH), 118.3, 117.7 (CH=CH$_2$), 114.3 (CH=CHCO), 69.7, 69.1 (CH$_2$CH=CH$_2$), 62.6 (CHN), 55.9 (OCH$_3$), 51.4 (CH$_2$Ph), 45.7 ((CH$_2$CO). HRMS: m/z (ESI+): 525.2398 (M + H$^+$). C$_{32}$H$_{33}$N$_2$O$_5$ requires 525.2389.

(R,S)-(E)-N-Benzyl-N-(1-(tert-butyl)-2-oxoazetidin-3-yl)-3-(4-(hydroxy)-3-methoxyphenyl)-acrylamide (**17a**). A solution of allyl derivative **16a** (84 mg, 187 μmol), Pd(PPh$_3$)$_2$Cl$_2$ (2.0 mg, 3.68 μmol, 0.015 eq. calculated on the number of allyl groups), and ammonium formate (18 mg, 280 μmol, 1.5 eq. calculated on the number of allyl groups) in dry CH$_3$CN (2.0 mL) in a pressure tube was first flushed with argon and then sealed and heated at 81 °C for 24 h. A tlc showed that the reaction was complete. The resulting dark solution was diluted with EtOAc and saturated aqueous NaHCO$_3$. The phases were separated and the aqueous one re-extracted three times with EtOAc. The organic phases were washed with brine and evaporated to dryness. Chromatography (PE/EtOAc 2:3 + 2% EtOH) gave pure **17a** as a white foam (71 mg, 93%). The purity by HPLC (for conditions see the general remarks) was >99%. R_f 0.30 (PE/EtOAc 50:50 + 2% EtOH). ^{1}H-NMR (CDCl$_3$): δ 7.70 (d, J = 15.3 Hz, 1 H, C*H*=CHCO), 7.41–7.25 (m, 5 H), 7.01 (d, J = 7.5 Hz, 1 H), 6.89 (s, 1 H), 6.88 (d, J not measurable, 1 H), 6.63 (d, J = 15.3 Hz, 1 H, CH=C*H*CO), 5.97 (s, 1 H, O*H*), 4.96 (broad s, 1 H, C*H*N), 4.84, 4.77 (AB syst., J = 17.3 Hz, 2 H, C*H*$_2$Ph), 3.87 (s, 3 H, OC*H*$_3$), 3.38 (t, J = 5.3 Hz., C*H*HCO), 3.18 (broad s, 1 H, CH*H*CO), 1.24 (s, 9 H, C(C*H*$_3$)$_3$). ^{13}C-NMR (CDCl$_3$): δ 167.8, 164.9 (C=O), 147.7, 146.7, 137.6, 127.5 (quat.), 144.3 (CH=CHCO), 128.9 (×2), 127.8, 126.7 (×2), 122.3, 114.8, 109.9 (aromatic CH), 114.2 (CH=CHCO), 61.3 (CHN), 56.0 (OCH$_3$), 53.2 (C(CH$_3$)$_3$), 51.0 (CH$_2$Ph), 43.7 ((CH$_2$CO), 27.4 (C(CH$_3$)$_3$). HRMS: m/z (ESI+): 409.2133 (M + H$^+$). C$_{24}$H$_{29}$N$_2$O$_4$ requires 409.2127.

(R,S)-(E)-N-Benzyl-3-(4-(hydroxy)-3-methoxyphenyl)-N-(1-(4-hydroxyphenyl)-2-oxoazetidin-3-yl)acrylamide (**17b**). It was prepared from **16b** (93 mg, 177 μmol), following the same procedure described above for **17a**, but in this case the workup was only the evaporation to dryness due to the low solubility of the product. Chromatography (DCM/MeOH 100:4) gave pure **17b** as a white solid (71 mg, 90%).

The purity by HPLC (for conditions see the general remarks) was 99%. R_f 0.42 (PE/ EtOAc 40:60 + 2% EtOH). ^{1}H-NMR (DMSO-d_6, 90 °C): δ 9.06 (s, 1 H, O*H*), 8.98 (s, 1 H, O*H*), (7.48 (d, J = 15.2 Hz, 1 H, C*H*=CHCO), 7.39–7.22 (m, 5 H), 7.21–7.14 (m, 3 H), 7.03 (dd, J = 8.2, 1.9 Hz, 1 H), 6.88 (d, J = 15.2 Hz, 1 H, CH=C*H*CO), 6.81–6.73 (m, 3 H), 5.26 (broad s, 1 H, C*H*N), 4.92, 4.77 (AB syst., J = 16.4 Hz, 2 H, C*H*$_2$Ph), 3.83 (t, J = 5.6 Hz, CH*H*CO), 3.79 (s, 3 H, OC*H*$_3$), 3.64–3.57 (m, 1 H, C*H*HCO). ^{13}C-NMR (DMSO-d_6, 90 °C): δ 166.7, 163.2 (C=O), 153.4, 148.8, 147.7, 138.5, 130.6, 126.2 (quat.), 143.3 (CH=CHCO), 128.6 (×2), 127.2, 126.6 (×2), 122.6, 117.4 (×2), 115.4 (×3), 111.2 (aromatic CH), 114.2 (CH=CHCO), 63.2 (CHN), 55.6 (OCH$_3$), 51.5 (CH$_2$Ph), 44.8 ((CH$_2$CO). HRMS: m/z (ESI+): 445.1766 (M + H$^+$). C$_{26}$H$_{25}$N$_2$O$_5$ requires 445.1763.

3.3. Thioflavin Experiments

One milliliter of DMSO was added to 1 mg of lyophilized synthetic peptide (Aβ1-42, AβpE3-42, AnaSpec, Fremont, CA, USA), to reach a final concentration of 1 mg mL^{-1}. Aliquots of 75 µL were lyophilized and stored at −20 °C until being used. For all experiments, stock peptides were reconstituted as reported [37]. For the preparation of the working samples, a stock solution of each peptide was divided into two or more aliquots. One was diluted to 5 µM in PBS containing 1% (v/v) DMSO to have a reference sample, and the others were diluted in PBS containing the appropriate quantity of polyphenol stock solution in DMSO in such a manner that each sample contains 1% of DMSO. The final pH was measured and eventually corrected to 7.4 using a few µL of 1 M HCl. Aβ peptides (5 µM) were incubated at 37 °C in the presence/absence of polyphenols as previously described and analyzed in parallel. ThT fluorescence was followed in time during aggregation. For this purpose, 47.5 µL of Aβ with and without test compounds were mixed with 2.5 µL ThT (400 µM) in a 3 mm path length fluorescence cuvette. ThT fluorescence was measured by using a luminescence spectrometer (LS50B, PerkinElmer, Waltham, MA, USA) at excitation and emission wavelengths of 440 nm (slit width = 5 nm) and 482 nm (slit width = 10 nm), respectively. ThT fluorescence data were plotted as a function of time and fitted by a sigmoidal curve described by the following equation: [38] $y = y_i + \frac{y_f - y_i}{1 + e^{(t-t_0)k_{fib}}}$ where y_i and y_f are the initial and final ThT fluorescence, respectively and k_{fib} is the fibril growing rate, t is time and t_0 is the time to 50% of maximal fluorescence. The lag time (T) is derived as $t_0 - 2/k_{fib}$.

4. Conclusions

In conclusion, the present study has demonstrated that replacement of the various pharmacophores in our previous leads may be critical, only small changes being permitted. A new possible hit has been selected from the second-generation library produced during this work. From the synthetic point of view, the feasibility of the overall strategy was again demonstrated, as well as the possibility to use the Ugi reaction with glycolaldehyde dimer for the synthesis of β-lactam-polyphenol hybrids. Due to the wide range of biological activities of polyphenols, we plan to investigate the many compounds synthesized through this strategy on other biological targets or in phenotypic assays. Studies towards this goal are in progress.

Supplementary Materials: The following are available online: copies of all NMR spectra.

Author Contributions: Conceptualization, L.B., C.D. and C.L.; Data curation, D.G., C.D., R.R. and C.L.; Formal analysis, G.B., C.L. and D.L.; Funding acquisition, L.B. and C.D.; Investigation, G.B., D.L., L.M. and C.L.; Methodology, D.G., A.B., L.M. and C.L.; Resources, L.B., A.B. and R.R.; Supervision, D.G., L.B., C.D. R.R. and C.L.; Writing—original draft, L.B. and C.L.; Writing—review & editing, D.G., C.D., L.M. and R.R.

Funding: This study is supported by Fondazione Cariplo, under the "Integrated Biotechnology and Bioeconomy" programme.

Acknowledgments: We thank Valeria Rocca for HPLC analyses and Andrea Armirotti for HRMS.

Conflicts of Interest: The authors declare no conflict of interest.

References

1. Abbas, M.; Saeed, F.; Anjum, F.M.; Afzaal, M.; Tufail, T.; Bashir, M.S.; Ishtiaq, A.; Hussain, S.; Suleria, H.A.R. Natural polyphenols: An overview. *Int. J. Food Prop.* **2017**, *20*, 1689–1699. [CrossRef]
2. Ganesan, K.; Xu, B.J. A critical review on polyphenols and health benefits of black soybeans. *Nutrients* **2017**, *9*, 17.
3. Quideau, S.; Deffieux, D.; Douat-Casassus, C.; Pouysegu, L. Plant polyphenols: Chemical properties, biological activities, and synthesis. *Angew. Chem. Int. Ed. Engl.* **2011**, *50*, 586–621. [CrossRef] [PubMed]
4. De Lucia, D.; Lucio, O.M.; Musio, B.; Bender, A.; Listing, M.; Dennhardt, S.; Koeberle, A.; Garscha, U.; Rizzo, R.; Manfredini, S.; et al. Design, synthesis and evaluation of semi-synthetic triazole-containing caffeic acid analogues as 5-lipoxygenase inhibitors. *Eur. J. Med. Chem.* **2015**, *101*, 573–583. [CrossRef] [PubMed]
5. Imai, K.; Nakanishi, I.; Ohkubo, K.; Ohba, Y.; Arai, T.; Mizuno, M.; Fukuzumi, S.; Matsumoto, K.; Fukuhara, K. Synthesis of methylated quercetin analogues for enhancement of radical-scavenging activity. *RSC Adv.* **2017**, *7*, 17968–17979. [CrossRef]
6. Khandelwal, A.; Hall, J.A.; Blagg, B.S.J. Synthesis and structure-activity relationships of EGCG analogues, a recently identified Hsp90 inhibitor. *J. Org. Chem.* **2013**, *78*, 7859–7884. [CrossRef] [PubMed]
7. Benchekroun, M.; Romero, A.; Egea, J.; Leon, R.; Michalska, P.; Buendia, I.; Jimeno, M.L.; Jun, D.; Janockova, J.; Sepsova, V.; et al. The antioxidant additive approach for Alzheimer's disease therapy: New ferulic (lipoic) acid plus melatonin modified tacrines as cholinesterases inhibitors, direct antioxidants, and nuclear factor (erythroid-derived 2)-like 2 activators. *J. Med. Chem.* **2016**, *59*, 9967–9973. [CrossRef] [PubMed]
8. Minassi, A.; Cicione, L.; Koeberle, A.; Bauer, J.; Laufer, S.; Werz, O.; Appendino, G. A multicomponent carba-betti strategy to alkylidene heterodimers—Total synthesis and structure-activity relationships of arzanol. *Eur. J. Org. Chem.* **2012**, 772–779. [CrossRef]
9. Montanari, S.; Bartolini, M.; Neviani, P.; Belluti, F.; Gobbi, S.; Pruccoli, L.; Tarozzi, A.; Falchi, F.; Andrisano, V.; Miszta, P.; et al. Multitarget strategy to address Alzheimer's disease: Design, synthesis, biological evaluation, and computational studies of coumarin-based derivatives. *ChemMedChem* **2016**, *11*, 1296–1308. [CrossRef]
10. Vo, D.D.; Elofsson, M. Synthesis of 4-formyl-2-arylbenzofuran derivatives by PdCl(C_3H_5)dppb-catalyzed tandem sonogashira coupling-cyclization under microwave irradiation—Application to the synthesis of viniferifuran analogues. *ChemistrySelect* **2017**, *2*, 6245–6248. [CrossRef]
11. Zhang, Z.; Su, P.; Li, X.; Song, T.; Chai, G.; Yu, X.; Zhang, K. Novel Mcl-1/Bcl-2 uual inhibitors created by the structure-based hybridization of drug-divided building blocks and a fragment deconstructed from a known two-face BH3 mimetic. *Arch. Pharm.* **2015**, *348*, 89–99. [CrossRef] [PubMed]
12. Tassano, E.; Alama, A.; Basso, A.; Dondo, G.; Galatini, A.; Riva, R.; Banfi, L. Conjugation of hydroxytyrosol with other natural phenolic fragments: From waste to antioxidants and antitumour compounds. *Eur. J. Org. Chem.* **2015**, 6710–6726. [CrossRef]
13. Espley, R.V.; Butts, C.A.; Laing, W.A.; Martell, S.; Smith, H.; McGhie, T.K.; Zhang, J.; Paturi, G.; Hedderley, D.; Bovy, A.; et al. Dietary flavonoids from modified apple reduce inflammation markers and modulate gut microbiota in mice. *J. Nutr.* **2014**, *144*, 146–154. [CrossRef] [PubMed]
14. Daglia, M.; Di Lorenzo, A.; Nabavi, S.F.; Talas, Z.S.; Nabavi, S.M. Polyphenols: Well beyond the antioxidant capacity: Gallic acid and related compounds as neuroprotective agents: You are what you eat! *Curr. Pharm. Biotech.* **2014**, *15*, 362–372. [CrossRef]
15. Fresco, P.; Borges, F.; Marques, M.P.M.; Diniz, C. The anticancer properties of dietary polyphenols and its relation with apoptosis. *Curr. Pharm. Design* **2010**, *16*, 114–134. [CrossRef]
16. Coppo, E.; Marchese, A. Antibacterial activity of polyphenols. *Curr. Pharm. Biotech.* **2014**, *15*, 380–390. [CrossRef]
17. Bahadoran, Z.; Mirmiran, P.; Azizi, F. Dietary polyphenols as potential nutraceuticals in management of diabetes: A review. *J. Diabetes Metab. Disord.* **2013**, *12*, 43. [CrossRef] [PubMed]
18. Ngoungoure, V.L.N.; Schluesener, J.; Moundipa, P.F.; Schluesener, H. Natural polyphenols binding to amyloid: A broad class of compounds to treat different human amyloid diseases. *Mol. Nutr. Food Res.* **2015**, *59*, 8–20. [CrossRef] [PubMed]
19. Liu, Y.; Wang, S.H.; Dong, S.Z.; Chang, P.; Jiang, Z.F. Structural characteristics of (−)-epigallocatechin-3-gallate inhibiting amyloid A beta 42 aggregation and remodeling amyloid fibers. *RSC Adv.* **2015**, *5*, 62402–62413. [CrossRef]

20. Pate, K.M.; Rogers, M.; Reed, J.W.; van der Munnik, N.; Vance, S.Z.; Moss, M.A. Anthoxanthin polyphenols attenuate a oligomer-induced neuronal responses associated with Alzheimer's disease. *CNS Neurosci. Therap.* **2017**, *23*, 135–144. [CrossRef]
21. Randino, R.; Grimaldi, M.; Persico, M.; De Santis, A.; Cini, E.; Cabri, W.; Riva, A.; D'Errico, G.; Fattorusso, C.; D'Ursi, A.M.; et al. Investigating the neuroprotective effects of turmeric extract: Structural interactions of beta-amyloid peptide with single curcuminoids. *Sci. Rep.* **2016**, *6*, 38846. [CrossRef] [PubMed]
22. Over, B.; Wetzel, S.; Gruetter, C.; Nakai, Y.; Renner, S.; Rauh, D.; Waldmann, H. Natural-product-derived fragments for fragment-based ligand discovery. *Nat. Chem.* **2013**, *5*, 21–28. [CrossRef] [PubMed]
23. Banfi, L.; Riva, R.; Basso, A. Coupling isocyanide-based multicomponent reactions with aliphatic or acyl nucleophilic substitution processes. *Synlett* **2010**, 23–41. [CrossRef]
24. Banfi, L.; Basso, A.; Lambruschini, C.; Moni, L.; Riva, R. Synthesis of seven-membered nitrogen heterocycles through the Ugi multicomponent reaction. *Chem. Heterocycl. Comp.* **2017**, *53*, 382–408. [CrossRef]
25. Lambruschini, C.; Galante, D.; Moni, L.; Ferraro, F.; Gancia, G.; Riva, R.; Traverso, A.; Banfi, L.; D'Arrigo, C. Multicomponent, fragment-based synthesis of polyphenol-containing peptidomimetics and their inhibiting activity on beta-amyloid oligomerization. *Org. Biomol. Chem.* **2017**, *15*, 9331–9351. [CrossRef] [PubMed]
26. Tomaselli, S.; Balducci, C.; Pagano, K.; Galante, D.; D'Arrigo, C.; Lambruschini, L.; Moni, L.; Banfi, L.; Molinari, H.; Forloni, G.; et al. Biophysical and in vivo studies identify a new natural-based polyphenol, counteracting Aβ oligomerization in vitro and Aβ oligomer-mediated memory impairment and neuroinflammation in an acute mouse model of Alzheimer's disease. *ACS Chem. Neurosci.* **2019**. submitted.
27. Banfi, L.; Basso, A.; Guanti, G.; Lecinska, P.; Riva, R. Multicomponent synthesis of benzoxazinones via tandem Ugi/Mitsunobu reactions: An unexpected cine-substitution. *Mol. Div.* **2008**, *12*, 187–190. [CrossRef]
28. Kim, Y.B.; Choi, E.H.; Keum, G.; Kang, S.B.; Lee, D.H.; Koh, H.Y.; Kim, Y.S. An efficient synthesis of morpholin-2-one derivatives using glycolaldehyde dimer by the Ugi multicomponent reaction. *Org. Lett.* **2001**, *3*, 4149–4152. [CrossRef]
29. Mossetti, R.; Pirali, T.; Tron, G.C. Synthesis of passerini-Ugi hybrids by a four-component reaction using the glycolaldehyde dimer. *J. Org. Chem.* **2009**, *74*, 4890–4892. [CrossRef]
30. Hanessian, S.; Couture, C.; Wiss, H. Design and reactivity of organic functional groups–Utility of imidazolylsulfonates in the synthesis of monobactams and 3-amino nocardicinic acid. *Can. J. Chem.* **1985**, *63*, 3613. [CrossRef]
31. Hanessian, S.; McNaughton-Smith, G.; Lombart, H.-G.; Lubell, W.D. Design and synthesis of conformationally constrained amino acids as versatile scaffolds and peptide mimetics. *Tetrahedron* **1997**, *53*, 12789–12854. [CrossRef]
32. Lambruschini, C.; Basso, A.; Moni, L.; Pinna, A.; Riva, R.; Banfi, L. Bicyclic heterocycles from levulinic acid through a fast and operationally simple diversity-oriented multicomponent approach. *Eur. J. Org. Chem.* **2018**, 5445–5455. [CrossRef]
33. Banfi, L.; Basso, A.; Guanti, G.; Kielland, N.; Repetto, C.; Riva, R. Ugi multicomponent reaction followed by an intramolecular nucleophilic substitution: Convergent multicomponent synthesis of 1-sulfonyl 1,4-diazepan-5-ones and of their benzo-fused derivatives. *J. Org. Chem.* **2007**, *72*, 2151–2160. [CrossRef] [PubMed]
34. Galante, D.; Ruggeri, F.S.; Dietler, G.; Pellistri, F.; Gatta, E.; Corsaro, A.; Florio, T.; Perico, A.; D'Arrigo, C. A critical concentration of N-terminal pyroglutamylated amyloid beta drives the misfolding of Ab1-42 into more toxic aggregates. *Int. J. Biochem. Cell Biol.* **2016**, *79*, 261–270. [CrossRef] [PubMed]
35. Ding, F.; Borreguero, J.M.; Buldyrev, S.V.; Stanley, H.E.; Dokholyan, N.V. Mechanism for the alpha-helix to beta-hairpin transition. *Proteins* **2003**, *53*, 220–228. [CrossRef] [PubMed]
36. Sanguinetti, M.; Sanfilippo, S.; Castagnolo, D.; Sanglard, D.; Posteraro, B.; Donzellini, G.; Botta, M. Novel macrocyclic amidinoureas: Potent non-azole antifungals active against wild-type and resistant Candida species. *Acs Med. Chem. Lett.* **2013**, *4*, 852–857. [CrossRef] [PubMed]
37. Galante, D.; Corsaro, A.; Florio, T.; Vella, S.; Pagano, A.; Sbrana, F.; Vassalli, M.; Perico, A.; D'Arrigo, C. Differential toxicity, conformation and morphology of typical initial aggregation states of A beta 1-42 and A beta py3-42 beta-amyloids. *Int. J. Biochem. Cell Biol.* **2012**, *44*, 2085–2093. [CrossRef]

38. Nielsen, L.; Khurana, R.; Coats, A.; Frokjaer, S.; Brange, J.; Vyas, S.; Uversky, V.N.; Fink, A.L. Effect of environmental factors on the kinetics of insulin fibril formation: Elucidation of the molecular mechanism. *Biochemistry* **2001**, *40*, 6036–6046. [CrossRef]

Sample Availability: No sample of compounds reported in this paper is available from the authors.

© 2019 by the authors. Licensee MDPI, Basel, Switzerland. This article is an open access article distributed under the terms and conditions of the Creative Commons Attribution (CC BY) license (http://creativecommons.org/licenses/by/4.0/).

Article

Isorhamnetin, Hispidulin, and Cirsimaritin Identified in *Tamarix ramosissima* Barks from Southern Xinjiang and Their Antioxidant and Antimicrobial Activities

Xiaopu Ren [1,2], Yingjie Bao [1], Yuxia Zhu [1], Shixin Liu [1], Zengqi Peng [1,*], Yawei Zhang [1,*] and Guanghong Zhou [1]

1. Key Laboratory of Meat Processing and Quality Control, Ministry of Education China, Jiangsu Collaborative Innovation Center of Meat Production and Processing, Quality and Safety Control, College of Food Science and Technology, Nanjing Agricultural University, Nanjing 210095, China; alarxp@126.com (X.R.); 2015208017@njau.edu.cn (Y.B.); 2016208019@njau.edu.cn (Y.Z.); liushixin_004@163.com (S.L.); ghzhou@njau.edu.cn (G.Z.)
2. Xinjiang Production & Construction Group Key Laboratory of Agricultural Products Processing in Xinjiang South, College of Life Science, Tarim University, Alar 843300, China
* Correspondence: zqpeng@njau.edu.cn (Z.P.); zhangyawei@njau.edu.cn (Y.Z.); Tel.: +86-25-84396558 (Z.P.)

Received: 15 December 2018; Accepted: 21 January 2019; Published: 22 January 2019

Abstract: As a natural potential resource, *Tamarix ramosissima* has been widely used as barbecue skewers for a good taste and unique flavor. The polyphenolics in the branch bark play a key role in the quality improvement. The purposes of the present work were to explore the polyphenolic composition of *T. ramosissima* bark extract and assess their potential antioxidant and antimicrobial activities. Hispidulin and cirsimaritin from *T. ramosissima* bark extract were first identified in the *Tamarix* genus reported with UPLC-MS analysis. Isorhamnetin (36.91 µg/mg extract), hispidulin (28.79 µg/mg extract) and cirsimaritin (13.35 µg/mg extract) are rich in the bark extract. The extract exhibited promising antioxidant activity with IC_{50} values of 117.05 µg/mL for 1,1-diphenyl-2-picrylhydrazyl (DPPH) and 151.57 µg/mL for hydroxyl radical scavenging activities, as well as excellent reducing power with an EC_{50} of 93.77 µg/mL. The bark extract showed appreciable antibacterial properties against foodborne pathogens. *Listeria monocytogenes* was the most sensitive microorganism with the lowest minimum inhibitory concentration (MIC) value of 5 mg/mL and minimum bactericidal concentration (MBC) value of 10 mg/mL followed by *S. castellani* and *S. aureus* among the tested bacteria. The *T. ramosissima* bark extract showed significantly stronger inhibitory activity against Gram-positive than Gram-negative bacteria. Nevertheless, this extract failed to show any activity against tested fungi. Overall, these results suggested that *T. ramosissima* shows potential in improving food quality due to its highly efficacious antioxidant and antibacterial properties.

Keywords: *Tamarix ramosissima*; polyphenolics; antioxidant activity; antimicrobial activity; isorhamnetin; hispidulin; cirsimaritin

1. Introduction

Various species of *Tamarix*, which are widely distributed throughout Europe, America, Asia, and Africa, have been used as herbal medicines in many civilizations due to the presence of polyphenolic compounds [1]. The methanolic extract of dried aerial components of *T. gallica* from India was found to prevent thioacetamide-promoted oxidative stress and toxicity and exhibited significant properties to reduce the susceptibility of the hepatic microsomal membrane to iron-ascorbate induced lipid peroxidation, H_2O_2 content, glutathione *S*-transferase, and xanthine oxidase activities in rats [2].

Yao et al. [3] showed that tamaractam, a new phenolic aromatic ring compound from *T. ramosissima* tender branches and leaves from the Ningxia province, displayed a strong inhibitory activity on cell proliferation in rheumatoid arthritis fibroblast-like synoviocytes, suggesting that it could remarkably induce cellular apoptosis and increase activated caspase-3/7 levels. Rahman et al [4] indicated that the methanolic extract of *T. indica* roots from Bangladesh exhibited excellent antinociceptive and anti-inflammatory properties. Significant writhing inhibition was produced by the extract in acetic acid-induced writhing in mice when comparable to the standard diclofenac sodium drug at the doses of 500 and 25 mg/kg body weight, respectively, and showed a significant anti-inflammatory activity against carrageenan-induced paw oedema in rats at oral doses of 200 and 400 mg/kg body weight compared to the standard drug aspirin. The wide spectrum of these medicinal properties may be mainly attributed to the presence of polyphenolic compounds in *Tamarix*, such as flavonoids and phenolic acids [1,2].

The leaves and flowers of *Tamarix* are rich in the polyphenolic compounds [5]. Sultanova et al. [1] identified tamarixetin in the leaves of *T. ramosissima* from southern Kazakhstan, and showed a high 1,1-diphenyl-2-picrylhydrazyl (DPPH) radical scavenging activity and antimicrobial activity against a number of pathogens, unambiguously specifying that the antioxidant and antibacterial activities of leaves were associated with the presence of polyphenolic compounds. Ksouri et al. [5] showed that syringic acid, isoquercetin and catechin were the major phenolics in the methanolic *T. gallica* leaf and flower extracts from south Tunis and that the flowers exhibited a higher antioxidant activity than that of the leaves, with IC_{50} values of the flower extracts being 1.3 (β-carotene bleaching) to 19-fold (lipid peroxidation inhibition) lower than those of leaves due to the higher total phenolic content (TPC). Meanwhile, the antibacterial properties of the leaf and flower methanolic extracts against human pathogen strains were also appreciable with a maximum inhibition zone of 15 mm against *Micrococcus luteus*. Unfortunately, there were few studies on the identification of polyphenolics, antioxidant and antimicrobial activities of the stem barks of *T. ramosissima* from southern Xinjiang.

T. ramosissima is one of the main constructive species in the Tarim River basin native to Northwestern China. In southern Xinjiang, for hundreds of years, green branches of *Tamarix* have been used as barbecue skewers for a good taste and unique flavor. And based on our previous study, the effective substances were largely concentrated in the bark of the green branches. In the present work, we aim to (i) identify and quantify the major polyphenols present in the green branches bark of *T. ramosissima*, and (ii) evaluate their antioxidant and antimicrobial activities against foodborne pathogens. We show that hispidulin and cirsimaritin are first identified in the *Tamarix* genus reported from *T. ramosissima* bark extract and the extract exhibits satisfying antioxidant and antimicrobial activities, which suggests *T. ramosissima* shows potential in improving food quality to promote health.

2. Results and Discussion

2.1. Content and Variety of Total Polyphenolics from T. ramosissima Barks

2.1.1. TPC and Total Flavonoid Content (TFC)

Many polyphenolic compounds, which are those containing one or more aromatic ring with one or more hydroxyl groups, act as antioxidants in natural plants due to their redox properties [5]. Flavonoids, which are one group of polyphenolics, are secondary metabolites in plants and act as antioxidants [6]. Many types of polyphenols, such as flavonoids and phenolic acids, were reported in *Tamarix* species [1,2,7–11]. In this work, the TPC of *T. ramosissima* bark extract was 323.45 mg gallic acid equivalent (GAE)/g, and the TFC was 87.32 quercetin equivalent (QE)/g. When TPC and TFC of the bark extract were compared with the data available for the same genus, it was found that the *T. ramosissima* bark contained much higher values, as shown in Table 1. The differences between the *Tamarix* species were obvious, and the TPC of *T. ramosissima* was 323.45 mg GAE/g extract, which was 9.39, 2.39 and 1.62 times higher than those of *T. gallica* leaves, *T. gallica* flowers, and *T. aphylla* leaves, respectively. The TFC of *T. ramosissima* in this work was much higher than those of *T. gallica* leaves and flowers. Additionally, the TPC and TFC varied greatly among different organs, and the bark was a standout organ based on Table 1.

Table 1. Comparison of TPC and TFC with published data of *Tamarix* family.

Tamarix Species	Tested Part	Location	TPC (mg GAE/g)	TFC (mg QE/g)	Reference
T. ramosissima	barks	South Xinjiang, China	323.45 ± 21.41	87.32 ± 1.65	This work
T. gallica	leaves	South Tunis	34.44 ± 3.40	3.91 ± 0.45	Ksouri et al. [5]
T. gallica	flowers	South Tunis	135.35 ± 7.70	12.33 ± 2.10	Ksouri et al. [5]
T. aphylla	leaves	South Algeria	199.54 ± 1.60	ND	Mohammedi [9]

Note: The TPC and TFC were presented as mean ± SD. ND means not detected.

2.1.2. Variety and Content of the Polyphenolics

UPLC-MS data of the bark extract are shown in Table 2 and Figure 1. A total of 13 polyphenolic compounds were identified and, for the first time, hispidulin and cirsimaritin were isolated from the genus *Tamarix*: they are active ingredients in a number of traditional Chinese herbs [12,13]. Regarding the other main polyphenolics, isorhamnetin was reported to have been identified from *T. hispida*, *T. elongata* and *T. laxa* which were all collected from southern Kazakhstan [11,14]. Quercetin was a common compound in the *Tamarix* species and had been identified by several researchers [5,8,10,11,15]. Compared to the results of this work, other researchers obtained different polyphenolic compounds from *Tamarix*. Sultanova et al. identified tamarixetin from *T. ramosissima* leaves in Kazakhstan [1] and isolated rhamnocitrin, isorhamnetin and a pentacyclic triterpenoid from the aerial components of *T. hispida* [14]. Ksouri et al. [5] identified polyphenolics from *T. gallica* leaves and flowers in Tunis, and his results showed that the flower polyphenolics consisted of seven phenolic acids (gallic, sinapic, chlorogenic, syringic, vanillic, *p*-coumaric, and *trans*-cinnamic acids), six flavonoids ((+)-catechin, isoquercetin, quercetin, apigenin, amentoflavone, and flavone), 12 phenolic compounds including gallic, sinapic, chlorogenic, syringic, vanillic, rosmarinic, *p*-coumaric, ferulic, and *trans*-cinnamic acids, as well as two flavonoids (quercetin and amentoflavone) which were identified from the leaves. Yao et al [3] identified only three compounds, tamaractam, *cis*-N-feruloyl-3-O-methyldopamine and *trans*-N-feruloyl-3-O-methyl- dopamine from *T. ramosissima* in Yinchuan, China.

Meanwhile, a further 4 polyphenolic compounds in high amounts, isorhamnetin, hispidulin, cirsimaritin, and quercetin, were quantified in this work (Table 3). However, no literature was found to quantify the concentration of polyphenolics from *Tamarix*.

Table 2. Peak identification of *T. ramosissima* bark extract using UPLC-MS.

Peak No.	Polyphenolic Compounds	Retention Time (min)	Empirical Formula	Calcd m/z	Obsd m/z [M + H]$^+$	Obsd m/z [M − H]$^+$
1	Quercetin 3-O-glucuronide	4.63	$C_{21}H_{18}O_{13}$	478.0747	479.0820	477.0676
2	Kaempferol 3-O-glucuronide	5.19	$C_{21}H_{18}O_{12}$	462.0798	463.0882	461.0802
3	Eriodictyol	6.63	$C_{15}H_{12}O_6$	288.0634	289.0703	287.0581
4	Quercetin	6.77	$C_{15}H_{10}O_7$	302.0427	303.0498	301.0404
5	Naringenin	7.65	$C_{15}H_{12}O_5$	272.0685	273.0753	271.0583
6	Tangeretin	7.72	$C_{20}H_{20}O_7$	372.1209	373.1274	371.1253
7	Kaempferol	7.85	$C_{15}H_{10}O_6$	286.0477	287.0547	285.0405
8	Hesperetin	8.07	$C_{16}H_{14}O_6$	302.0790	303.0856	301.0718
9	Isorhamnetin	9.25	$C_{16}H_{12}O_7$	316.0583	317.0653	315.0510
10	Hispidulin	10.41	$C_{16}H_{12}O_6$	300.0634	301.0705	299.0562
11	Apigenin	10.42	$C_{15}H_{10}O_5$	270.0528	271.0597	269.0427
12	Glycitein	11.90	$C_{16}H_{12}O_5$	284.0685	285.0754	283.0664
13	Cirsimaritin	11.91	$C_{17}H_{14}O_6$	314.0790	315.0863	313.0717

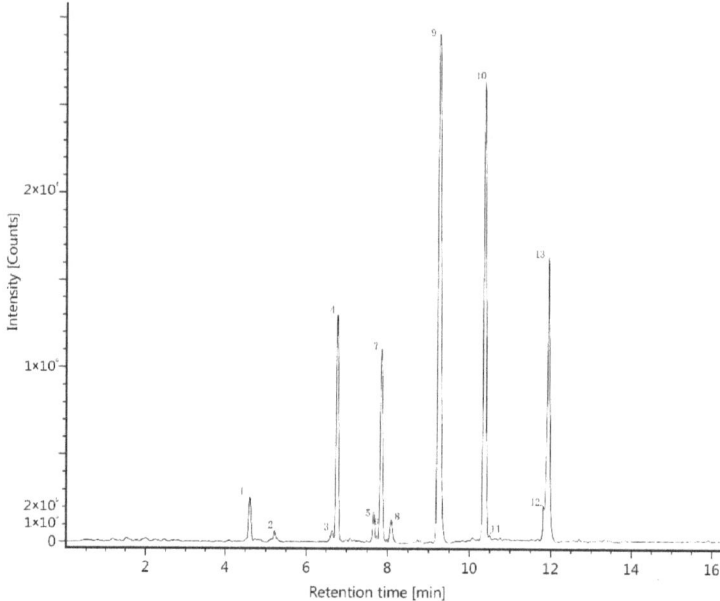

Figure 1. UPLC-MS chromatogram of polyphenolics in *T. ramosissima* crude extract infusion: quercetin 3-O-glucuronide (1), kaempferol 3-O-glucuronide (2), eriodictyol (3), quercetin (4), naringenin (5), tangeretin (6), kaempferol (7), hesperetin (8), isorhamnetin (9), hispidulin (10), apigenin (11), glycitein (12), cirsimaritin (13).

Table 3. Concentrations of four representative polyphenolic standards.

Compounds	Concentration (µg/mg)
Isorhamnetin	36.9055
Hispidulin	28.7915
Cirsimaritin	13.3513
Quercetin	4.2065

2.2. Antioxidant Activity of the Bark Extract

2.2.1. DPPH Scavenging Activity

The DPPH free radicals have been extensively used to investigate the scavenging activity of natural antioxidants. Figure 2A shows the results of the scavenging DPPH radical activity of *T. ramosissima* bark extracts. The scavenging activity increased sharply when the concentration increased from 25 to 200 µg/mL and trended towards a plateau after 300 µg/mL (91.70% ± 0.78), at which maximal scavenging activity on the DPPH radicals was reached. There was no significant difference of the scavenging activity between the bark extract groups and the control (ascorbic acid) ($p > 0.05$) at concentrations of more than 200 µg/mL. There was also no significant difference among the concentrations of 300, 400, and 500 µg/mL of the bark extract groups ($p > 0.05$). The IC_{50} value (117.05 µg/mL) of *T. ramosissima* bark extract for scavenging activity against DPPH was much higher than the results of Ksouri et al. [5], who found that the IC_{50} values on the DPPH radical of the *T. gallica* flower and leaf extracts were 2 and 9 µg/mL, respectively, due to the structural conformation of the antioxidants. In the main polyphenols of the *T. gallica* flower and leaf extracts, the second hydroxyl group of gallic, chlorogenic, and rosmarinic acid and catechin were in the ortho position, while the compounds of sinapic, syringic, vanillic, and ferulic acid had ortho-methoxy substitutions. These compounds in the flower and leaf extracts included a second hydroxyl group in the ortho or para position, and the ortho-methoxy substitution group increased antioxidant efficiencies [16]. However, isorhamnetin and hispidulin in the bark extract had the second hydroxyl group in the meta position. This is potentially the reason for the lower scavenging activity on the DPPH radicals compared to the results of Ksouri et al. [5].

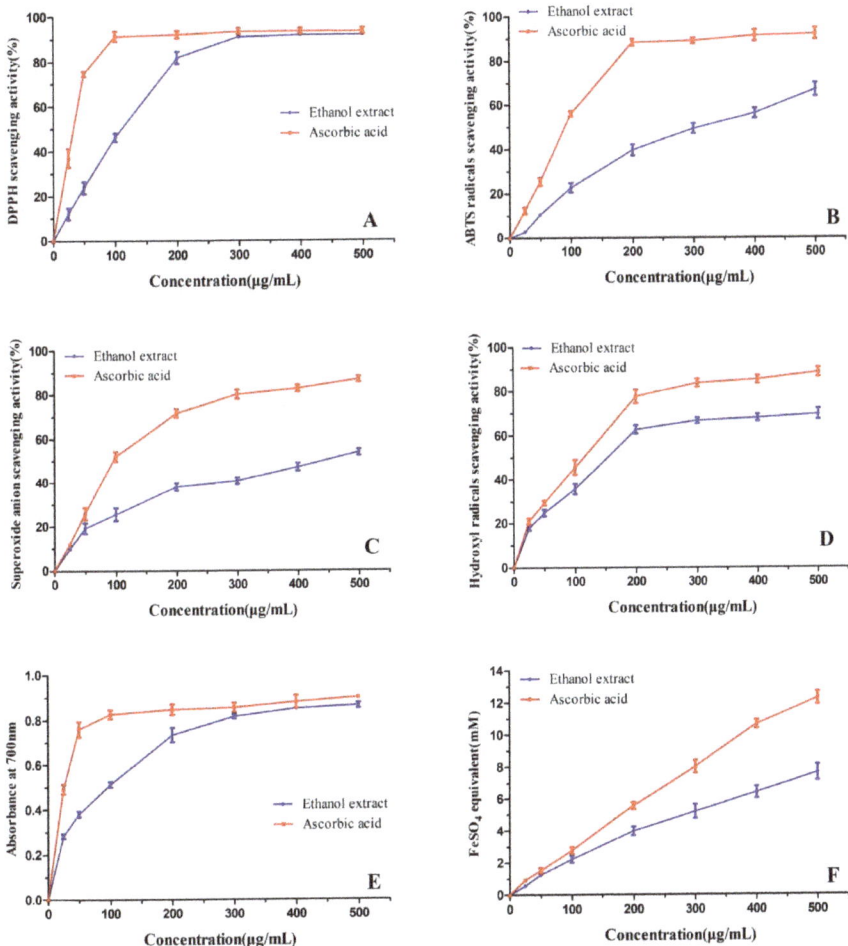

Figure 2. Scavenging activities on 1,1-diphenyl-2-picrylhydrazyl (DPPH) (**A**), 2,2′-Azinobis-(3-ethylbenzthiazoline-6-sulphonate) (ABTS) (**B**), superoxide radicals (**C**), hydroxyl radicals (**D**), reducing power (**E**) and ferric reducing antioxidant power (FRAP) (**F**) of the bark extract of *T. ramosissima* and ascorbic acid. Data are presented as means ± SD of triplicates.

2.2.2. 2,2′-Azinobis-(3-ethylbenzthiazoline-6-sulphonate) (ABTS) Scavenging Activity

The scavenging activity of 2,2′-Azinobis-(3-ethylbenzthiazoline-6-sulphonate) (ABTS) radicals is another widely used method to assess the radical scavenging capacity of natural antioxidants [17]. As shown in Figure 2B, the ABTS radical scavenging activity of the *T. ramosissima* bark extract increased with the sample concentration. The scavenging activity of the polyphenolics on the ABTS· was correlated with their concentrations [18]. At a concentration of 500 μg/mL, the ABTS radical scavenging activity of the bark extract was 67.02%. The scavenging activities of bark extract groups from 25 to 500 μg/mL were significant lower than those of the control ($p < 0.05$). Additionally, there were significant differences in the ABTS scavenging activity among the bark extract concentrations ($p < 0.05$). The IC_{50} value of the bark extract on the ABTS radicals (314.88 µg/mL) was considerably higher than that of ascorbic acid (82.90 µg/mL), similar to the IC_{50} of the *T. gallica* flower extract (316.7 µg/mL), and 3.3 times lower than that of the *T. gallica* leaf extract [5]. The decreased IC_{50} of the *T. ramosissima* bark extract and *T. gallica* flower extract were probably because of the increased level of TPC and TFC (Table 1). Moreover, the difference in polyphenolic components among the extracts may be one of the reasons for the different scavenging activity on the ABTS radicals.

2.2.3. Superoxide Anion Scavenging Activity

According to Yagi [19], as opposed to the mechanism of DPPH and ABTS radical reactions, the mechanism of superoxide and hydroxyl radicals was peroxide decomposition. The results of scavenging activity on superoxide radicals of the bark extract are presented in Figure 2C. The scavenging activity on the superoxide radical of the bark extract increased for concentrations ranging from 0 to 500 µg/mL. At the concentrations lower than 50 µg/mL of bark extract, there was no significant difference in scavenging activity of superoxide radicals between the bark extract and the control group. The difference in scavenging activity on superoxide radical between the concentrations of 200 and 300 µg/mL of bark extract was not significant. The IC_{50} value of the bark extract for the superoxide radical was 442.53 µg/mL, much higher than that of the results of Ksouri et al. [5], who found that the IC_{50} on the scavenging activity of the *T. gallica* flower and leaf extracts were 3 and 22 µg/mL, respectively. In the *T. gallica* flower and leaf extracts, hydroxyl groups were located at the 3'- and 4'-positions of the B-ring in quercetin, catechin, and isoquercetin that exhibited much higher scavenging activity on superoxide and hydroxyl radicals [19,20]. In the present work, isorhamnetin, hispidulin, and cirsimaritin in the bark extract possessed only one single hydroxyl group at the 4'-position of the B-ring, which exhibited slight scavenging activity.

2.2.4. Hydroxyl Radicals Scavenging Activity

As Figure 2D shows, the scavenging activities on hydroxyl radicals increased with the concentrations. The hydroxyl radical scavenging activity increased rapidly in the range of 25 to 200 µg/mL, before reaching a plateau from 300 to 500 µg/mL. Significant differences in the activity were found among the concentrations of 25, 50, 100, and 200 µg/mL of the bark extracts ($p < 0.05$). The scavenging activities of the bark extract in these concentrations were not significantly different from the ascorbic acid ($p > 0.05$). The IC_{50} of the bark extract on the hydroxyl radical was 151.57 µg/mL, which was not significantly higher than that of the ascorbic acid (114.08 µg/mL), suggesting a satisfactory hydroxyl radical scavenging activity. It was evident from Figure 2C,D that the scavenging activity of the bark extract on hydroxyl radicals was stronger than on superoxide radicals. At 200 µg/mL, the bark extract quenched 62.66% of hydroxyl radicals while only quenching 38.28% of superoxide radicals. The stronger scavenging activity on hydroxyl radicals of the bark extract than on superoxide radicals may be associated with the carbonyl function at the C-4 position in the structures of isorhamnetin, hispidulin, and cirsimaritin. The result was in accordance with Husain et al., who found that the presence of a carbonyl functional group at the C-4 position played an important role in the hydroxyl radical quenching ability and that quercetin possessing a carbonyl showed higher quenching ability on the hydroxyl radicals than the carbonyl-devoid catechin [20].

2.2.5. Reducing Power

The reducing power of natural antioxidants, which was determined using a modified Fe^{3+} to Fe^{2+} reduction assay, has been declared to be associated with their antioxidant activity [21]. As shown in Figure 2E, the reducing power of *T. ramosissima* bark extract was excellent and increased with the amount of the extract. The reducing power of bark extract at concentrations below 200 µg/mL increased rapidly with concentrations but was significantly lower compared with the control ($p < 0.05$). At the concentration of 300 µg/mL, the reducing power of the bark extract was similar to that of ascorbic acid ($p > 0.05$). Furthermore, there were also no significant differences among the concentrations of 300, 400, and 500 µg/mL of the bark extract groups ($p > 0.05$). The EC_{50} of the bark extract was 93.77 µg/mL, similar to the EC_{50} (76.67 µg/mL) of *T. gallica* leaf extract [5]. However, in another report of Ksouri et al. [22] on *T. gallica* flower and leaf extracts from a different location, the EC_{50} values of the reducing power were 84.3 and 205 µg/mL, respectively.

2.2.6. Ferric Reducing Antioxidant Power (FRAP)

The mechanism of the ferric reducing antioxidant power (FRAP) assay was similar to the reducing power, both of which relied on the ability of antioxidants to reduce iron (III) to iron (II). More specifically, in the FRAP assay, a ferric tripyridyltriazine (FeIII-TPTZ) complex was reduced to the ferrous form (FeII-TPTZ) at low pH value, and several researchers considered the assay as the total antioxidant power [23,24]. In the present work, the trends for ferric ion reducing activities of *T. ramosissima* bark extract and ascorbic acid are shown in Figure 2F. For both of them, the $FeSO_4$ equivalent (mM/g) clearly increased due to the formation of the Fe^{2+}-TPTZ complex with increasing concentration. The $FeSO_4$ equivalent increased linearly with the concentrations ($R^2 = 0.9879$ for the bark extract, 0.9961 for ascorbic acid). There was no significant difference between the bark extract group and the control at concentrations ranging from 0 to 100 µg/mL ($p > 0.05$). However, the differences became larger with increasing concentrations. At the concentration of 500 µg/mL, the $FeSO_4$ equivalent was 7.64 mM/g for the bark extract and 12.28 mM/g for the positive control, both of which were higher than those of *Barringtonia racemosa* and kudingcha crude extracts determined by the same methods [25,26]. Hidalgo et al. [27] found that in the FRAP assay, the polyphenols with 3-hydroxyl group in the C-ring (such as isorhamnetin and quercetin in the bark extract of the present work) showed high antioxidant power and that the antioxidant activity did not decrease when the 3-hydroxyl group in the C-ring was blocked if the 3′,4′-dihydroxy structure in the B-ring was retained (as in the case of quercetin 3-O-glucuronide in the bark extract of the present work).

As previously mentioned, the antioxidant activity of the bark extract of *T. ramosissima* was satisfactory and the presence of polyphenolics, especially flavonoids, probably played a significant role. The findings of Ksouri et al. [5] largely supported the claims of the present work, which suggests that the extracts of *T. gallica* showed very high antioxidant activity due to the presence of polyphenolics and that the antioxidant properties exhibit a high positive correlation with polyphenolic content. Sultanova et al. [1] also specifically stated that the antioxidant activity was associated with the presence of polyphenolic substances. In the bark extract of *T. ramosissima*, different polyphenols exhibited various antioxidant activities. Isorhamnetin, a 3′-O-methylated metabolite of quercetin, exerts excellent antioxidant effects, which reduces oxidative stress due to free radicals by induction of NF-E2-related factor 2 (Nrf2)-dependent antioxidant genes [28]. Cirsimaritin is a small molecular natural flavonoid, mainly derived from the medicinal plant Herba Artemisiae Scopariae, which exhibits a variety of beneficial activities including antioxidant activity [29]. Quercetin is the typical flavonoid structure and is widely used as a nutritional supplement due to its antioxidant and anti-inflammatory properties [30]. Hispidulin, also named 6-methoxy-5,7,4′-trihydroxyflavone, has been shown to possess anti-inflammatory and antioxidative activities [31], although the antioxidant activity of hispidulin is very weak compared with quercetin [32]. All of these polyphenolic compounds contributed to the antioxidant activity of the *T. ramosissima* bark extract. In general, the antioxidant property of a given compound is thought to be closely linked to its structural features, including the ortho-dihydroxy structure in the B-ring, the 2,3 double bond in conjugation with a 4-oxo function, the presence of the 3- and 5-OH functions and glycosidic moieties and the number and position of hydroxyl and methoxy groups [27]. Moreover, the nature of the radical and its specific reaction mechanism also exert great influence on the antioxidant activity of the tested polyphenolics. All of these elements determine the final effects of the polyphenolics.

2.3. Antimicrobial Activity

The antimicrobial activities of *T. ramosissima* bark extract at different concentrations are shown in Table 4. For the Gram-positive bacteria, the inhibition zone against *Listeria monocytogenes* at the concentration of 10 mg/mL of the bark extract was the largest ($p < 0.05$). The inhibition zones against *L. monocytogenes* and *Staphylococcus aureus* were larger than that of *Bacillus cereus* at the concentration of 5 mg/mL, and significantly increased with the concentrations ($p < 0.05$). However, the inhibition zone against *S. aureus* was not influenced by the concentration of the bark extract. Among these Gram-positive strains, *L. monocytogenes* was the most sensitive microorganism with the lowest minimum bactericidal concentration (MBC) values of 10 mg/mL. Excellent antibacterial activity against Gram-positive bacteria was also supported by Ksouri et al. [5], who found that the inhibition zone ranged from 7.00 to 15.00 mm at the concentrations of 2, 4 and 100 mg/mL of the *T. gallica* leaf and flower extracts. For *S. aureus*, the bark extract in this work showed much larger inhibition zones compared with the *T. gallica* leaf and flower extracts. The methanol extract of *T. indica* from Bangladesh also exhibited similar antibacterial activity against *S. aureus* with an inhibition zone of 10.80 mm [4]. Nevertheless, the *T. ramosissima* leaf extract from southern Kazakhstan showed no antibacterial activity against *S. aureus* and *B. cereus* [1]. The inhibitory effects of the bark extract on bacterial pathogens could be attributed to the isorhamnetin, hispidulin, or cirsimaritin in the *T. ramosissima* bark extract, of which the purified hispidulin showed antibacterial activity against *B. subtilis* with minimum inhibitory concentration (MIC) values of 50 μg/mL and *S. aureus* with MIC values of 100 μg/mL [33].

Table 4. Antibacterial activity of *T. ramosissima* bark extracts at different concentrations. Minimum bactericidal concentration (MBC), minimum inhibitory concentration (MIC).

Bacterial Strains		The Extract Concentrations (mg/mL)	Diameter of Inhibition Zones		MIC (mg/mL)	MBC (mg/mL)
			Bark Extract	Gentamycin (10 UI)		
Gram-positive strains	*Staphylococcus aureus*	1	10.60 ± 0.54 A			
		5	11.08 ± 0.71 A	20.90 ± 0.14	5	15
		10	11.16 ± 0.91 B			
	Listeria monocytogenes	1	10.14 ± 0.48 cAB			
		5	11.18 ± 0.21 bA	20.30 ± 0.42	5	10
		10	12.26 ± 0.61 aA			
	Bacillus cereus	1	9.53 ± 0.62 bBC			
		5	9.72 ± 0.16 bB	18.00 ± 0.00	5	20
		10	10.88 ± 0.30 aB			
Gram-negative strains	*Escherichia coli*	1	9.26 ± 0.65 bC			
		5	9.58 ± 0.34 bB	17.80 ± 0.28	10	25
		10	10.68 ± 0.81 aB			
	Pseudomonas aeruginosa	1	8.20 ± 0.51 bD			
		5	8.85 ± 0.70 abC	17.10 ± 0.14	>10	NA
		10	9.38 ± 0.28 aC			
	Salmonella typhimurium	1	8.00 ± 0.24 bD			
		5	8.34 ± 0.66 bC	21.75 ± 0.35	>10	NA
		10	9.48 ± 0.52 aC			
	Shigella castellani	1	10.65 ± 0.45 A			
		5	10.74 ± 0.50 A	20.90 ± 0.14	5	15
		10	11.16 ± 0.69 B			

Note: Inhibition zone calculated in diameter around the disc (mean ± SD). Different lowercase letters (a~c) within the same bacteria mean significant differences between different concentrations ($p < 0.05$); Different capital letters (A–D) within the same concentration mean significant differences between different bacteria ($p < 0.05$) and no letters indicates no significant difference ($p > 0.05$). NA represented not active. The diameter of disc was 6 mm. Each experiment was done in triplicate.

The inhibition zones against Gram-negative bacteria were increased with the concentrations ($p < 0.05$), except in the case of *Shigella castellani*. In the Gram-negative bacteria, the inhibition zones against *S. castellani* were the largest ($p < 0.05$) at the concentrations of 1 or 5 mg/mL of the bark extract. The inhibition zones against *Escherichia coli* and *S. castellani* were larger than those of *Pseudomonas aeruginosa* and *Salmonella typhimurium* at the concentration of 10 mg/mL ($p < 0.05$). *S. castellani* was the most sensitive to the bark extract, with the lowest MIC of 5 mg/mL and MBC values of 15 mg/mL. These results were consistent with those of Rahman et al., who found that the *T. indica* extract exhibited the largest inhibition zones against *S. sonnie* [4]. In the present work, *E. coli* was the second most sensitive to the bark extract with a MIC of 10 mg/mL and MBC values of 25 mg/mL. The *T. gallica* leaf and flower extracts from Tunis also had inhibitory effects on *E. coli* [5].

Compared to Gram-negative bacteria, Gram-positive bacteria were sensitive to the bark extract, with the mean of inhibition zones significantly larger at the concentrations of 1, 5 and 10 mg/mL ($p < 0.05$). Similar results can also be observed from the MIC values of these bacteria. These differences in inhibitory effects of the bark extract between Gram-negative bacteria and Gram-positive bacteria may be associated with the different cell wall components of the bacteria [34]. The antibacterial activity of the *T. ramosissima* bark extract was lower than that of gentamycin ($p < 0.05$). The *T. ramosissima* bark extract failed to show any activity against four tested fungi. The results of this work suggested that there may be a huge potential application of *T. ramosissima* to prevent the growth of foodborne pathogens.

3. Materials and Methods

3.1. Plant Materials and Extraction Procedure

Fresh twigs of *T. ramosissima* were collected from the Tarim River basin on the edge of the Taklamakan desert in Xinjiang (40°25′28.81″ N, 81°14′18.14″ E, Southwest of Alar city) in June 2017. The plant was identified by an expert taxonomist, Mr. Pang, who had been engaged on taxonomic research of *Tamarix* in South Xinjiang for a significant length of time at Tarim University. The peeled bark off green branches were air-dried and milled to a fine powder (mesh size: 1 mm). Approximately 500 g of powdered barks were soaked in 2500 mL of 60% ethanol for 40 min. Ultrasonic-assisted extraction was used, and the power was 600 W (VCX 750, Sonics, USA). The extracts were filtered through a Whatman no. 4 filter paper and evaporated under vacuum. Then, the vacuum freeze dryer (LGJ-10C, Four-ring, China) was employed to obtain the ethanol extracts and they were stored at -20 °C until analysis. All solvents and reagents were of analytical grade.

3.2. Determination of Total Polyphenolic Content

The TPC in the *T. ramosissima* bark extract was determined according to the Folin–Ciocalteu procedure as described by Jayasinghe et al [35] with some modifications, using gallic acid as a standard. 0.05 g of the bark extract was dissolved in 2 mL of ethanol-water solution and mixed with 2 mL of Folin–Ciocalteu's phenol reagent. The mixture was kept for 3 min at room temperature, and then 2 mL of Na_2CO_3 solution (10%, w/v) was added to the mixture. The reaction mixture was allowed 1 h to incubate at ambient temperature in the dark, and the absorbance was then read at 530 nm. Deionized water was used as a blank sample. A standard calibration curve of gallic acid (0–0.4 mg/mL) was plotted ($R^2 = 0.9995$). The results were expressed as milligrams of gallic acid equivalent per gram of the bark extract (mg GAE/g). All analyses were performed in triplicate.

3.3. Determination of Total Flavonoid Content

The TFC in the *T. ramosissima* bark extract was determined using the aluminum chloride colorimetric method as described by Chang et al [36] with some modifications using quercetin as a standard. In this experiment, 0.05 g of the bark extract was dissolved in 2 mL of 80% ethanol and then separately mixed with 0.1 mL of 10% aluminum chloride, 0.1 mL of 1 M potassium acetate and 2.8 mL deionized water. The reaction mixture was incubated at room temperature for 30 min, and the absorbance was then measured at 415 nm. Deionized water was used as a blank sample. A standard calibration curve of quercetin (0–0.1 mg/mL) was plotted ($R^2 = 0.9991$). The concentration of flavonoid was expressed as milligrams of quercetin equivalent per gram of extract (mg QE/g). All analyses were performed in triplicate.

3.4. Analysis of Bark Extract by UPLC-MS

The samples were analyzed using a UPLC-MS following a method previously published, with some modifications [37]. The samples were analyzed by an LC-MS system (G2-XS QTof, Waters). First, 2 µL solutions were injected into the UPLC column (2.1 mm × 100 mm ACQUITY UPLC BEH C18 column containing 1.7 µm particles) with a flow rate of 0.4 mL/min. Buffer A consisted of 0.1% formic acid in water, and buffer B consisted of 0.1% formic acid in acetonitrile. The gradient was 5% Buffer B for 0.5 min, 5–95% Buffer B for 11 min, and 95% Buffer B for 2 min. Mass spectrometry was performed using an electrospray source in negative ion mode with the MSe acquisition mode and a selected mass range of 50–1200 m/z. The lock mass option was enabled using leucine-enkephalin (m/z 554.2615) for recalibration. The ionization parameters were the following: capillary voltage was 2.5 kV, collision energy was 40 eV, source temperature was 120 °C, and desolvation gas temperature was 400 °C. For quantification purposes, the four major compounds, isorhamnetin, hispidulin, cirsimaritin, and quercetin (Sigma, USA) were quantified using the calibration curves of their corresponding standards. Data acquisition and processing were performed using Masslynx 4.1 and results were presented as µg/mg extract.

3.5. Determination of Antioxidant Assays

3.5.1. DPPH Radical Scavenging Activity

The DPPH free radical scavenging activity of the *T. ramosissima* bark extract was conducted according to the method of Yang et al [38], with some modifications. 0.5 mL of the bark extract solution at various concentrations (25, 50, 100, 200, 300, 400, and 500 µg/mL) was mixed with 3.5 mL of freshly prepared DPPH-ethanol solution (1×10^{-4} mol/L). The reaction mixture was mixed vigorously for 15 s and then kept in the dark for 30 min at room temperature. The absorbance was then measured at 517 nm, and the DPPH free radical scavenging percentage was calculated based on the following equation:

$$\text{DPPH radical scavenging activity (\%)} = [1 - (A_1 - A_2)/A_0] \times 100 \qquad (1)$$

where A_1 was the Abs of the sample, A_2 was the Abs of the sample only (ethanol instead of DPPH), and A_0 was the Abs of the control (deionized water instead of sample solution). The bark extract concentration that inhibited 50% of the DPPH radicals (IC$_{50}$) was calculated and expressed as µg/mL. The ascorbic acid was used as a positive control and conducted in parallel. The experiment was carried out in triplicate.

3.5.2. ABTS Radical Scavenging Activity

The ABTS free radical-scavenging activity of the *T. ramosissima* bark extract was determined according to Liu et al. [26]. A mixture (1:1, v/v) of ABTS (7.0 mM) and potassium persulfate (4.95 mM) was incubated at 25 °C overnight in the dark to prepare the fresh stock solution. A working solution was prepared by diluting the stock solution with phosphate buffer solution (pH 7.4, 0.2 M) to obtain an absorbance of 0.70 ± 0.02 at 734 nm. Then, 20 µL of the bark extract solutions at various concentrations (25, 50, 100, 200, 300, 400, and 500 µg/mL) were mixed with 200 µL working solution and kept in the dark for 30 min. The absorbance was then measured at 734 nm. The ascorbic acid was used as a positive control, and the ABTS radical scavenging activity was calculated according to the following equation:

$$\text{ABTS radical scavenging activity (\%)} = [1 - (A_1 - A_2)/A_0] \times 100 \qquad (2)$$

where A_0, A_1 and A_2 have the same meaning as in Equation (1). The IC$_{50}$ value was also calculated and expressed as µg/mL. All the analyses were performed in triplicate.

3.5.3. Superoxide Anion Radical Scavenging Activity

The superoxide anion radical scavenging of the *T. ramosissima* bark extract was evaluated based on the method of Liu et al. [39], with slight modifications. The reaction mixture contained 1 mL of 50 µM nitro-blue tetrazolium (NBT), 1 mL of 78 µM nicotinamide adenine dinucleotide (NADH), 1 mL of 10 µM phenazine methosulfate (PMS), and 1 mL of the bark extract at different concentrations (25, 50, 100, 200, 300, 400 and 500 µg/mL). After incubation in the dark for 10 min at 37 °C, the absorbance was measured at 560 nm. The ascorbic acid was used as a positive control and triplicate tests were conducted for each sample. The superoxide anion radical scavenging activity was calculated according to Equation (3):

$$\text{Superoxide anion radical scavenging activity (\%)} = [1 - (A_1 - A_2)/A_0] \times 100 \quad (3)$$

where A_0, A_1 and A_2 have the same meaning as in Equation (1). The IC_{50} value was calculated and expressed as µg/mL.

3.5.4. Hydroxyl Radical Scavenging Activity

The hydroxyl radical scavenging activity of the *T. ramosissima* bark extract was assayed according to Li [40], with some modifications. The hydroxyl radicals were generated in a Fenton reaction by incubating for 60 min at 37 °C in the presence of 1.0 mM $FeSO_4$, 2.0 mM H_2O_2, 1.0 mM Ethylene diamine tetraacetic acid (EDTA), 1.0 mM sodium salicylate, 20 mM NaH_2PO_4-Na_2HPO_4 buffer (pH 7.4), and sample solutions of various concentrations (25, 50, 100, 200, 300, 400, and 500 µg/mL) in a final volume of 5 mL. The solutions of $FeSO_4$ and H_2O_2 were freshly produced in distilled water just before use. After incubation, the absorbance was measured at 532 nm and all samples were determined in three replicates. The results were calculated according to Equation (4):

$$\text{Hydroxyl radical scavenging activity (\%)} = [1 - (A_1 - A_2)/A_0] \times 100 \quad (4)$$

where A_0, A_1 and A_2 have the same meaning as in Equation (1). The IC_{50} value was calculated and expressed as µg/mL.

3.5.5. Reducing Power

The reducing power of *T. ramosissima* bark extract was determined according to the method of Ye et al [41] with several modifications. The sample was dissolved in phosphate buffer saline (PBS) (pH 6.6, 0.2 M) to afford various concentrations (25, 50, 100, 200, 300, 400, and 500 µg/mL). 2.5 mL sample solutions and 2.5 mL of 1% potassium ferricyanide were mixed and incubated at 50 °C for 20 min. Then, the mixture was cooled to 25 °C and 2.5 mL of 10% trichloroacetic acid was added. The mixture was centrifuged at $650 \times g$ for 10 min. The supernatant (2.5 mL) was mixed with 2.5 mL of distilled water and 0.5 mL of 0.1% ferric chloride. After thoroughly mixing, the absorbance was measured at 700 nm by a microplate reader, and the ascorbic acid was used as a positive control. Triplicate tests were conducted for each sample, and a higher absorbance indicated a higher reducing power. The reducing power was calculated according to the formula below:

$$\text{Reducing power} = A_1 - A_2 \quad (5)$$

where A_1 is the absorbance of the sample and A_2 is the absorbance of the sample only (distilled water instead of ferric chloride). The EC_{50} value is the effective dose of the bark extract yielding an absorbance of 0.5 for reducing power and expressed as µg/mL.

3.5.6. Ferric Reducing Antioxidant Power (FRAP) Assay

The FRAP assay was performed according to Liu et al. [26], with some modifications. Briefly, 0.2 mL *T. ramosissima* bark extract solution at different concentrations (25, 50, 100, 200, 300, 400, and 500 µg/mL) was added to 3.8 mL of FRAP reagent (10 volumes of 300 mM sodium acetate buffer at pH 3.6, 1 volumes of 10.0 mM 2,4,6-tripyridyl-s-triazine (TPTZ) solution, and 1 volume of 20.0 mM $FeCl_3 \cdot 6H_2O$ solution), and then the mixture was warmed to 37 °C for 30 min in the dark. The absorbance was then measured at 593 nm. The antioxidant activity was calculated from the calibration curve (y = 0.1446x + 0.0478) in the range from 0.15 to 1.50 mM $FeSO_4$ with good linearity (R^2 = 0.9943). The results were expressed as mM $FeSO_4$ equivalent/g extract. The ascorbic acid was used as a positive control, and all analyses were performed in triplicate.

3.6. Determination of Antimicrobial Activity

3.6.1. Microorganisms

The target bacterial strains used in the study were seven American Type of Culture Collection (ATCC) strains: Gram-positive bacteria including *S. aureus* (ATCC 25923), *L. monocytogenes* (ATCC 13932), *B. cereus* (ATCC 11778), and Gram-negative bacteria including *E. coli* (ATCC 35218), *P. aeruginosa* (ATCC 27853), *S. typhimurium* (ATCC 14028), and *S. castellani* (ATCC 12022). The strains were first grown in Mueller Hinton (MH, Hopebio, Qingdao, China) broth at 37 °C for 24 h prior to seeding onto the MH agar.

For the antifungal activity, four clinical isolates of *Penicillium expansum*, *Aspergillus niger*, *Acremonium strictum*, and *Penicillium citrinum* were first grown on Sabouraud dextrose agar (SDA, HANGWEI, China) plates at 28 °C for 48 h.

3.6.2. Agar Disc Diffusion Method

The Kirby–Bauer agar disc diffusion method was performed to determine the antibacterial and antifungal activity of the *T. ramosissima* bark extract following procedures previously described by Ebani et al [42]. Briefly, isolated colonies were selected to prepare the bacterial inocula in sterile saline solutions to obtain a turbidity equivalent to a 0.5 McFarland standard, approximately 1 to 2×10^7 CFU/mL. Meanwhile, the fungal spores were picked into the sterile saline solution and adjusted to 10^4 to 10^5 CFU/mL. An aliquot of 0.1 mL of bacterial (or spore) suspension was spread onto an MH agar plate (SDA for fungi) and then a sterile filter disc with 6 mm diameter (Whatman paper no. 3), which was impregnated with the bark extract of different concentrations (1, 5, 10 mg/mL) in dimethyl sulfoxide (DMSO, Oxoid), was placed on the surface of the media. The plates were incubated at 37 °C for 24 h, and 28 °C for 48 h for fungi, followed by the measurement of the diameter of the growth inhibition zone expressed in millimeters (mm). DMSO was used as negative control, with gentamycin and ketoconazole (Amresco, USA) as positive controls for bacteria and fungi, respectively. All tests were performed in triplicate.

3.6.3. Determination of Minimum Inhibitory Concentration (MIC)

The MIC of the *T. ramosissima* bark extract was determined for each bacterial strain, which was sensitive to the bark extract of the Kirby-Bauer assay, using the microdilution broth method in 96-well microplates. The tests were carried out in MH broth, and a stock solution (20 mg/mL) of the bark extract was prepared in 10% DMSO. An aliquot of this solution was serially diluted (two-fold) with MH broth to obtain a concentration ranging from 20 to 0.15625 mg/mL. After careful mixing, the microplates were incubated at 37 °C for 24 h. The absorbances of the plates at 620 nm were measured with a multiplate reader (SpectraMax M3, Molecular Devices, USA). The MIC value was defined as the lowest concentration of the bark extract at which there was no visible growth of the microorganisms [43]. The results were expressed as mg/mL, and all tests were performed in triplicate.

3.6.4. Determination of Minimum Bactericidal Concentration (MBC)

The MBC was determined by adding 100 µL aliquots of the microplate well contents which did not show any growth in the MIC test, to the MH agar and then incubating at 37 °C for 24 h. The MBC was defined as the lowest concentration of the bark extract which showed no bacterial growth [43].

3.7. Statistical Analysis

All analyses were performed in triplicate. The results were expressed as means ± standard deviations. The IC_{50} values were calculated by linear regression analysis. The statistics analysis was performed using SPSS software (version 22.0) by one-way analysis of variance (ANOVA). p-value < 0.05 was regarded as significant.

4. Conclusions

In the current work, hispidulin and cirsimaritin were first identified from the bark extract of *T. ramosissima* in southern Xinjiang. Hispidulin, cirsimaritin, and isorhamnetin, abundant polyphenolics in the extract, were found at levels of 28.79, 13.35 and 36.91 µg/mg extract, respectively. The bark extract showed satisfying antioxidant activity with a similar DPPH scavenging activity and reducing power to ascorbic acid at 300, 400, and 500 µg/mL. The bark extract exhibited excellent antibacterial activities against foodborne pathogens. *L. monocytogenes*, *S. aureus* and *B. cereus* were sensitive to the bark extract compared with *E. coli*, *P. aeruginosa*, *S. typhimurium* and *S. castellani*. *L. monocytogenes* was the most sensitive bacteria among these foodborne pathogens. Future investigation should focus on the effects of *T. ramosissima* on hazardous substance formation and sensory properties during food processing, as well as the potential health-promoting properties.

Author Contributions: Conceptualization, X.R., Z.P. and Y.Z.; Funding acquisition, Z.P.; Investigation, X.R.; Methodology, X.R., Y.B. and Y.Z.; Project administration, Z.P. and Y.Z.; Resources, X.Z.; Software, S.L.; Writing—original draft, X.R.; Writing—review and editing, Z.P. and G.Z.

Funding: This research was funded by the National Key R&D Program of China (NO. 2018YFD0502306) and the Open Research Fund for Young Teachers in Nanjing Agricultural University and Tarim University (NO. TDNNLH201701).

Conflicts of Interest: The authors declare no conflict of interest.

References

1. Sultanova, N.; Makhmoor, T.; Abilov, Z.A.; Parween, Z.; Omurkamzinova, V.B.; Atta ur, R.; Choudhary, M.I. Antioxidant and antimicrobial activities of *Tamarix ramosissima*. *J. Ethnopharmacol.* **2001**, *78*, 201–205. [CrossRef]
2. Sehrawat, A.; Sultana, S. *Tamarix gallica* ameliorates thioacetamide-induced hepatic oxidative stress and hyperproliferative response in Wistar rats. *J. Enzym. Inhib. Med. Chem.* **2006**, *21*, 215–223. [CrossRef]
3. Yao, Y.; Jiang, C.S.; Sun, N.; Li, W.Q.; Niu, Y.; Han, H.Q.; Miao, Z.H.; Zhao, X.X.; Zhao, J.; Li, J. Tamaractam, a new bioactive Lactam from *Tamarix ramosissima*, induces apoptosis in rheumatoid arthritis fibroblast-like synoviocytes. *Molecules* **2017**, *22*, 96. [CrossRef]
4. Rahman, M.A.; Haque, E.; Hasanuzzaman, M.; Shahid, I.Z. Antinociceptive, Antiinflammatory and antibacterial properties of *Tamarix indica* roots. *Int. J. Pharmacol.* **2011**, *7*, 527–531. [CrossRef]
5. Ksouri, R.; Falleh, H.; Megdiche, W.; Trabelsi, N.; Mhamdi, B.; Chaieb, K.; Bakrouf, A.; Magne, C.; Abdelly, C. Antioxidant and antimicrobial activities of the edible medicinal halophyte *Tamarix gallica* L. and related polyphenolic constituents. *Food Chem. Toxicol.* **2009**, *47*, 2083–2091. [CrossRef] [PubMed]
6. Lee, K.G.; Shibamoto, T.; Takeoka, G.R.; Lee, S.E.; Kim, J.H.; Park, B.S. Inhibitory effects of plant-derived flavonoids and phenolic acids on malonaldehyde formation from ethyl arachidonate. *J. Agric. Food Chem.* **2003**, *51*, 7203–7207. [CrossRef]
7. Abouzid, S.F.; Ali, S.A.; Choudhary, M.I. A new ferulic acid ester and other constituents from *Tamarix nilotica* leaves. *Chem. Pharm. Bull.* **2009**, *57*, 740–742. [CrossRef]

8. Bikbulatova, T.N.; Korul'kina, L.M. Composition of *Tamarix hokenakeri* and *T. ramosissima*. *Chem. Nat. Compd.* **2001**, *37*, 216–218. [CrossRef]
9. Mohammedi, Z.; Atik, F. Impact of solvent extraction type on total polyphenols content and biological activity from *Tamarix aphylla* (L.) Karst. *Int. J. Pharm. Biol. Sci.* **2011**, *2*, 609–615.
10. Parmar, V.S.; Bisht, K.S.; Sharma, S.K.; Jain, R.; Taneja, P.; Singh, S.; Simonsen, O.; Boll, P.M. Highly oxygenated bioactive flavones from *Tamarix*. *Phytochemistry* **1994**, *36*, 507–511. [CrossRef]
11. Umbetova, A.K.; Choudhary, M.I.; Sultanova, N.A.; Burasheva, G.S.; Abilov, Z.A. Flavonoids of plants from the genus *Tamarix*. *Chem. Nat. Compd.* **2005**, *41*, 728–729. [CrossRef]
12. Gao, H.; Wang, H.; Peng, J. Hispidulin induces apoptosis through mitochondrial dysfunction and inhibition of P13k/Akt signalling pathway in HepG2 cancer cells. *Cell Biochem. Biophys.* **2014**, *69*, 27–34. [CrossRef] [PubMed]
13. Yan, H.; Wang, H.; Ma, L.; Ma, X.; Yin, J.; Wu, S.; Huang, H.; Li, Y. Cirsimaritin inhibits influenza A virus replication by downregulating the NF-κB signal transduction pathway. *Virol. J.* **2018**, *15*, 88. [CrossRef] [PubMed]
14. Sultanova, N.; Makhmoor, T.; Yasin, A.; Abilov, Z.A.; Omurkamzinova, V.B.; Atta ur, R.; Choudhary, M.I. Isotamarixen—A new antioxidant and prolyl endopeptidase-inhibiting triterpenoid from *Tamarix hispida*. *Planta Med.* **2004**, *70*, 65–67. [CrossRef]
15. Ishak, M.S.; El-Sissi, H.I.; Nawwar, M.A.; El-Sherbieny, A.E. Tannins and polyphenolics of the galls of *Tamarix aphylla* I. *Planta Med.* **1972**, *21*, 246–253. [CrossRef] [PubMed]
16. Brand-Williams, W.; Cuvelier, M.E.; Berset, C. Use of a free radical method to evaluate antioxidant activity. *LWT Food Sci. Technol.* **1995**, *28*, 25–30. [CrossRef]
17. Xie, M.; Hu, B.; Wang, Y.; Zeng, X. Grafting of gallic acid onto chitosan enhances antioxidant activities and alters rheological properties of the copolymer. *J. Agric. Food Chem.* **2014**, *62*, 9128–9136. [CrossRef] [PubMed]
18. Oszmianski, J.; Wojdylo, A.; Lamer-Zarawska, E.; Swiader, K. Antioxidant tannins from Rosaceae plant roots. *Food Chem.* **2007**, *100*, 579–583. [CrossRef]
19. Yagi, K. A rapid method for evaluation of autoxidation and antioxidants. *Agric. Biol. Chem.* **1970**, *34*, 142–145. [CrossRef]
20. Husain, S.R.; Cillard, J.; Cillard, P. Hydroxyl radical scavenging activity of flavonoids. *Phytochemistry* **1987**, *26*, 2489–2491. [CrossRef]
21. Juntachote, T.; Berghofer, E. Antioxidative properties and stability of ethanolic extracts of Holy basil and Galangal. *Food Chem.* **2005**, *92*, 193–202. [CrossRef]
22. Ksouri, R.; Megdiche, W.; Falleh, H.; Trabelsi, N.; Boulaaba, M.; Smaoui, A.; Abdelly, C. Influence of biological, environmental and technical factors on phenolic content and antioxidant activities of Tunisian halophytes. *C. R. Biol.* **2008**, *331*, 865–873. [CrossRef] [PubMed]
23. Benzie, I.F.F.; Strain, J.J. The ferric reducing ability of plasma (FRAP) as a measure of "antioxidant power": The FRAP assay. *Anal. Biochem.* **1996**, *239*, 70–76. [CrossRef] [PubMed]
24. Benzie, I.F.F.; Strain, J.J. Ferric reducing antioxidant power assay: Direct measure of total antioxidant activity of biological fluids and modified version for simultaneous measurement of total antioxidant power and ascorbic acid concentration. *Methods Enzymol.* **1999**, *299*, 15–27. [CrossRef] [PubMed]
25. Kong, K.W.; Mat-Junit, S.; Aminudin, N.; Ismail, A.; Abdul-Aziz, A. Antioxidant activities and polyphenolics from the shoots of *Barringtonia racemosa* (L.) Spreng in a polar to apolar medium system. *Food Chem.* **2012**, *134*, 324–332. [CrossRef]
26. Liu, L.; Sun, Y.; Laura, T.; Liang, X.; Ye, H.; Zeng, X. Determination of polyphenolic content and antioxidant activity of kudingcha made from *Ilex kudingcha* C.J. Tseng. *Food Chem.* **2009**, *112*, 35–41. [CrossRef]
27. Hidalgo, M.; Sanchez-Moreno, C.; de Pascual-Teresa, S. Flavonoid-flavonoid interaction and its effect on their antioxidant activity. *Food Chem.* **2010**, *121*, 691–696. [CrossRef]
28. Seo, S.; Seo, K.; Ki, S.H.; Shin, S.M. Isorhamnetin inhibits reactive oxygen species-dependent hypoxia inducible factor (HIF)-1α accumulation. *Biol. Pharm. Bull.* **2016**, *39*, 1830–1838. [CrossRef]
29. Quan, Z.; Gu, J.; Dong, P.; Lu, J.; Wu, X.; Wu, W.; Fei, X.; Li, S.; Wang, Y.; Wang, J.; et al. Reactive oxygen species-mediated endoplasmic reticulum stress and mitochondrial dysfunction contribute to cirsimaritin-induced apoptosis in human gallbladder carcinoma GBC-SD cells. *Cancer Lett.* **2010**, *295*, 252–259. [CrossRef]

30. Kim, B.H.; Choi, J.S.; Yi, E.H.; Lee, J.K.; Won, C.; Ye, S.K.; Kim, M.H. Relative antioxidant activities of quercetin and its structurally related substances and their effects on NF-κB/CRE/AP-1 signaling in murine macrophages. *Mol. Cells* **2013**, *35*, 410–420. [CrossRef]
31. He, L.; Wu, Y.; Lin, L.; Wang, J.; Wu, Y.; Chen, Y.; Yi, Z.; Liu, M.; Pang, X. Hispidulin, a small flavonoid molecule, suppresses the angiogenesis and growth of human pancreatic cancer by targeting vascular endothelial growth factor receptor 2-mediated PI3K/Akt/mTOR signaling pathway. *Cancer Sci.* **2011**, *102*, 219–225. [CrossRef] [PubMed]
32. Chen, Y.T.; Zheng, R.L.; Jia, Z.J.; Ju, Y. Flavonoids as superoxide scavengers and antioxidants. *Free Radic. Biol. Med.* **1990**, *9*, 19–21. [CrossRef]
33. Osman, W.J.A.; Mothana, R.A.; Basudan, O.; Mohammed, M.S.; Mohamed, M.S. Antibacterial effect and radical scavenging activity of hispidulin and nepetin; a two flvaones from *Tarconanthus camphoratus* L. *World J. Pharm. Res.* **2015**, *4*, 424–433.
34. Scalbert, A. Antimicrobial properties of tannins. *Phytochemistry* **1991**, *30*, 3875–3883. [CrossRef]
35. Jayasinghe, C.; Gotoh, N.; Aoki, T.; Wada, S. Phenolics composition and antioxidant activity of sweet basil (*Ocimum basilicum* L.). *J. Agric. Food Chem.* **2003**, *51*, 4442–4449. [CrossRef] [PubMed]
36. Chang, C.C.; Yang, M.H.; Wen, H.M.; Chern, J.C. Estimation of total flavonoid content in propolis by two complementary colorimetric methods. *J. Food Drug Anal.* **2002**, *10*, 178–182.
37. Esteban-Fernandez, A.; Zorraquin-Pena, I.; Ferrer, M.D.; Mira, A.; Bartolome, B.; Gonzalez de Llano, D.; Victoria Moreno-Arribas, M. Inhibition of oral pathogens adhesion to human gingival fibroblasts by wine polyphenols alone and in combination with an oral probiotic. *J. Agric. Food Chem.* **2018**, *66*, 2071–2082. [CrossRef]
38. Yang, L.C.; Li, R.; Tan, J.; Jiang, Z.T. Polyphenolics composition of the leaves of *Zanthoxylum bungeanum* maxim. grown in Hebei, China, and their radical scavenging activities. *J. Agric. Food Chem.* **2013**, *61*, 1772–1778. [CrossRef]
39. Liu, F.; Ooi, V.E.C.; Chang, S.T. Free radical scavenging activities of mushroom polysaccharide extracts. *Life Sci.* **1997**, *60*, 763–771. [CrossRef]
40. Li, H.; Wang, Q.J. Evaluation of free hydroxyl radical scavenging activities of some Chinese herbs by capillary zone electrophoresis with amperometric detection. *Anal. Bioanal. Chem.* **2004**, *378*, 1801–1805. [CrossRef]
41. Ye, Z.; Wang, W.; Yuan, Q.; Ye, H.; Sun, Y.; Zhang, H.; Zeng, X. Box-Behnken design for extraction optimization, characterization and in vitro antioxidant activity of *Cicer arietinum* L. hull polysaccharides. *Carbohydr. Polym.* **2016**, *147*, 354–364. [CrossRef] [PubMed]
42. Ebani, V.V.; Nardoni, S.; Bertelloni, F.; Giovanelli, S.; Rocchigiani, G.; Pistelli, L.; Mancianti, F. Antibacterial and antifungal activity of essential oils against some pathogenic bacteria and yeasts shed from poultry. *Flavour Frag. J.* **2016**, *31*, 302–309. [CrossRef]
43. Bassanetti, I.; Carcelli, M.; Buschini, A.; Montalbano, S.; Leonardi, G.; Pelagatti, P.; Tosi, G.; Massi, P.; Fiorentini, L.; Rogolino, D. Investigation of antibacterial activity of new classes of essential oils derivatives. *Food Control* **2017**, *73*, 606–612. [CrossRef]

Sample Availability: Samples of the compounds (isorhamnetin, hispidulin, cirsimaritin, and quercetin) are not available from the authors.

© 2019 by the authors. Licensee MDPI, Basel, Switzerland. This article is an open access article distributed under the terms and conditions of the Creative Commons Attribution (CC BY) license (http://creativecommons.org/licenses/by/4.0/).

Article

Melaleuca styphelioides Sm. Polyphenols Modulate Interferon Gamma/Histamine-Induced Inflammation in Human NCTC 2544 Keratinocytes

Ferdaous Albouchi [1,2], Rosanna Avola [3], Gianluigi Maria Lo Dico [4], Vittorio Calabrese [5], Adriana Carol Eleonora Graziano [3], Manef Abderrabba [1] and Venera Cardile [3,*]

- [1] Laboratoire Matériaux-Molécules et Applications, University of Carthage, IPEST, B.P. 51 2070, La Marsa, Tunisia; ferdaous.albouchi@gmail.com (F.A.); abderrabbamanef@gmail.com (M.A.)
- [2] Faculte des Sciences de Bizerte, University of Carthage, Jarzouna, 7021, Bizerte, Tunisia
- [3] Department of Biomedical and Biotechnological Sciences, Section of Physiology, University of Catania, Via Santa Sofia, 97-95123 Catania, Italy; rosanna.avola@unict.it (R.A.); acegraz@unict.it (A.C.E.G.)
- [4] Istituto Zooprofilattico Sperimentale della Sicilia "A. Mirri", Via Gino Marinuzzi 3, 90129 Palermo, Italy; gigilodico@gmail.com
- [5] Department of Biomed & Biotech Sciences, School of Medicine, University of Catania, Via Santa Sofia 97, 95125 Catania, Italy; calabres@unict.it
- * Correspondence: cardile@unict.it; Tel.: +39-095-4781318

Received: 10 September 2018; Accepted: 29 September 2018; Published: 2 October 2018

Abstract: *Melaleuca styphelioides*, known as the prickly-leaf tea tree, contains a variety of bioactive compounds. The purposes of this study were to characterize the polyphenols extracted from *Melaleuca styphelioides* leaves and assess their potential antioxidant and anti-inflammatory effects. The polyphenol extracts were prepared by maceration with solvents of increasing polarity. The LC/MS-MS technique was used to identify and quantify the phenolic compounds. An assessment of the radical scavenging activity of all extracts was performed using 2,2-diphenyl-1-picrylhydrazyl (DPPH), 2,2'-azinobis-(3-ethylbenzothiazoline-6-sulphonate) (ABTS$^+$), and ferric reducing antioxidant power (FRAP) assays. The anti-inflammatory activity was determined on interferon gamma (IFN-γ)/histamine (H)-stimulated human NCTC 2544 keratinocytes by Western blot and RT-PCR. Compared to other solvents, methanolic extract presented the highest level of phenolic contents. The most frequent phenolic compounds were quercetin, followed by gallic acid and ellagic acid. DPPH, ABTS$^+$, and FRAP assays showed that methanolic extract exhibits strong concentration-dependent antioxidant activity. IFN-γ/H treatment of human NCTC 2544 keratinocytes induced the secretion of high levels of the pro-inflammatory mediator inter-cellular adhesion molecule-1 (ICAM-1), nitric oxide synthase (iNOS), cyclooxygenase-2 (COX-2), and nuclear factor kappa B (NF-κB), which were inhibited by extract. In conclusion, the extract of *Melaleuca styphelioides* leaves is rich in flavonoids, and presents antioxidant and anti-inflammatory proprieties. It can be proposed as a useful compound to treat inflammatory skin diseases.

Keywords: *Melaleuca styphelioides*; polyphenols; LC/MS-MS; anti-oxidant activity; anti-inflammatory activity; keratinocytes

1. Introduction

Inflammation is a complicated series of protective responses, involving several cell types and various putative modulators and mediators [1]. Skin provides both chemical and physical barriers and contains the cellular components of a rapid innate immune response [2], which protects the organism against external invasions.

Keratinocytes are the major constituents of the epidermis and appear to be exposed to stressful conditions such as toxins and microbes. They are also the major players of the complex response in the skin, conducting the activation of a diversity of Toll-like receptors [3,4]. Histamine (H) and interferon gamma (IFN-γ) can induce inflammatory responses in keratinocytes, highlighted by the activation of the pro-inflammatory mediators, such as the nuclear factor kappa B (NF-κB), inter-cellular adhesion molecule-1 (ICAM-1), nitric oxide synthase (iNOS), and cyclooxygenase-2 (COX-2) [5–7]. Inhibitors of these mediators are extensively used to treat several inflammatory diseases. However, the majority of such anti-inflammatory drugs (steroidal or non-steroidal molecules) are highly toxic, and their use is often associated with harmful effects on the gastrointestinal tract, such as mucosal lesions, bleeding, and peptic ulcers [8].

Currently, a regular growing interest in plant polyphenols is proposed as an alternative to treat skin inflammatory diseases. Plant polyphenols are one of the greatest groups, with the largest chemical diversity. In addition, polyphenols are widely used in traditional medicine to treat several skin diseases, like vitiligo, psoriasis [9], and atopic dermatitis [10], as well as to accelerate skin wound healing [9]. However, the mechanisms by which plant phenolic compounds exert their anti-inflammatory effect are poorly understood. The potential anti-inflammatory effects of polyphenols have been attributed to either their free radical scavenging activities or classical chain-breaking, antioxidant, and transition metal chelating activity [11]. Other in-vitro studies using skin cells suggest that polyphenols inhibit the activation of cellular functions by multiple mechanisms, such as the modulation of intracellular signal transduction and transcription of a number of genes, direct interaction with several receptors, and post-translational modulation of enzymatic activities [12,13]. It is well-documented that polyphenols inhibit major pro-inflammatory skin enzymes, such as COX, LOX, iNOS, NADPH oxidase, NF-κB, ICAM-1, and PLA2 [12]. The anti-inflammatory mechanism of plant polyphenols depends on various factors, such as chemical structure, synergy of other phenols, cell types, and inductor used [12].

Melaleuca (M.) *styphelioides* Sm. belongs to the *Melaleuca* genus, commonly known as the prickly-leaf tea tree. The *Melaleuca* genus, or tea tree, is presented by approximately 260 described species native to Australia and is widespread in Southeast Asia, the Caribbean, and the Southern United States. In different parts of world, tea tree is used in traditional medicine as a treatment for insect bites, bruises, skin infections, flu and colds, acne vulgaris, psoriasis, inflammation, dermatitis, an antimicrobial agent, and as an insect repellant. Moreover, *Melaleuca* species are used in the manufacture of cosmetics, such as shampoos, soaps, and some cosmetic products [14]. *M. styphelioides* is well-known species that has been reported to have the strongest production of scented essential oil and tannins, as well as amount of flavonoid compounds and phenolic acids [15]. Several studies have reported the antioxidant, antimicrobial, hepatoprotective, and anti-proliferative activities of the essential oil and extract isolated from *M. styphelioides* leaves [15–18].

Here, we aim to (i) characterize and quantify the polyphenol compounds present in *M. styphelioides* leaves, and (ii) evaluate their antioxidant and anti-inflammatory activities in IFN-γ/H-induced inflammation of human NCTC 2544 keratinocytes. To our knowledge, this is the first report that demonstrates the antioxidant and anti-inflammatory activities of polyphenols extracted by *M. styphelioides* leaves.

2. Results and Discussion

In this work, four solvents of increasing polarities were chosen for the evaluation of phenolic compounds contents of *M. styphelioides* leaves, namely hexane, diethyl ether, ethyl acetate, and methanol.

The yields of total phenol, flavonoids, and tannins from *M. styphelioides* leaf extracts are shown in Table 1. The maximum content of phenols total (PT) (142.7 ± 3.15 mg gallic acid equivalents (GAE)/g dry extract) and flavonoid total (FT) (31.54 ± 1.99 mg quercetin equivalents (QE)/g dry extract) was obtained in the MeOH extract. Lower amounts were found in the ethyl acetate and Et$_2$O extracts, and even lower amounts in the hexane extract. The highest amount of TC (19.9 ± 2.9 mg Eq Catéchine/g dry extract) was determined in the EtOAc and MeOH extracts (15.2 ± 1.9 mg Eq Catéchine/g dry extract), while no TC was present in Et$_2$O and hexane extracts (Table 1). The important phenolic compounds extraction yields, found in the MeOH extract, has been attributed to its high solubility, low toxicity, medium polarity, and high extraction capacity. Our results are in agreement with the previously report of Al-Sayed et al. [15], where the authors report a higher amount of phenolic and flavonoid components in the MeOH extract of *M. styphelioides* leaves.

Table 1. Total phenols, flavonoids, and tannins content in *M. styphelioides* leaves.

M. styphelioides Extracts	Total Phenolic mg GAE/g Dry Extract	Total Flavonoid mg QE/g Dry Extract	Total Tannins mg Eq Catéchine/g Dry Extract
E. MeOH	142.7 ± 3.15	31.54 ± 1.99	15.2 ± 1.9
E. EtOAc	97.39 ± 7.69	26.8 ± 2.4	19.9 ± 2.9
E. Et$_2$O	22.95 ± 0.4	7.83 ± 1.11	4.1 ± 1.3
E. Hex	3.27 ± 2.1	nd	nd

nd: not determined.

Based on the optimization condition of LC/MS-MS, the MeOH extract of *M. styphelioides* leaves was subjected to identification and quantification of the polyphenol components, in order to better discuss its biological potential. The detailed phenolic composition was determined using our standard library information (peak retention time, [M − H$^-$] (m^2), and LC-MS/MS data). *M. styphelioides* leaf extract showed the presence of several types of phenols belonging to diverse phenolic families, such as phenolic acids and flavonoids. Fifteen phenolic compounds were identified and quantified, with a range of 3.04–5.61 min retention times (RT) in negative polarity mode (Figure 1 and Table 2). Qualitatively, the phenolic profile included seven phenolic acids (vanillic acid, gallic acid, caffeic acid, syringic acid, chlorogenic acid, ferulic acid, and ellagic acid), nine flavonoids (apigenin, kaempferol, myricetin, naringenin, quercetin, luteolin, pinocembrin, hesperidin, and rutin). In quantitative terms, *M. styphelioides* polyphenol extract contained a rich source of bioactive phenols, mainly phenolic acids (25%) and flavonoids (70.6%). Most of the detected flavonoids corresponded to quercetin (53.99%) and apigenin (4%), followed by kaempferol (3.2%). The major phenolic acid compound found in metanolic extract was gallic acid (13.5%), followed by ellagic acid (6.3%) and vanillic acid (4.3%).

Table 2. Phenolic composition of M. styphelioides methanolic extract by LC/MS-MS.

C.A.S. Number	RT (min)	Mass (amu) [M − H$^-$]	Fragments (m/z)	Compounds	Phenolic Family	Concentration (µg/kg DW)
327-97-9	0.59	353.8	191.20	Chlorgenic Acid	Phenolic acids	36 ± 7
149-91-7	1.28	169.01	125.00	Gallic Acid	Phenolic acids	1116 ± 127
1135-24-6	3.04	193.05	143.00	Ferulic Acid	Phenolic acids	86 ± 15
331-39-5	3.24	179.03	135.02	Caffeic Acid	Phenolic acids	92 ± 17
530-57-4	3.51	197.04	121.00	Syringic Acid	Phenolic acids	292 ± 35
207671-50-9	3.56	610.01	300.30	Rutin	Flavonoids	259 ± 31
520-26-3	3.59	609.20	301.00	Hesperidina	Flavonoids	177 ± 27
121-34-6	3.73	167.04	108.00	Vanillic Acid	Phenolic acids	359 ± 36
491-70-3	4.36	285.04	133.00	Luteolin	Flavonoids	56 ± 10
476-66-4	4.40	302.20	131.98	Ellagic Acid	Phenolic acids	522 ± 47
529-44-2	4.46	317.04	151.00	Myricetin	Flavonoids	160 ± 24
117-39-5	5.06	447.09	151.00	Quercetin	Flavonoids	4440 ± 355
67604-48-2	5.55	272.06	119.00	Naringenin	Flavonoids	23 ± 5
520-36-5	5.61	271.08	117.00	Apigenin	Flavonoids	336 ± 38
520-18-3	6.49	285.04	108.00	Kaempferol	Flavonoids	271 ± 35

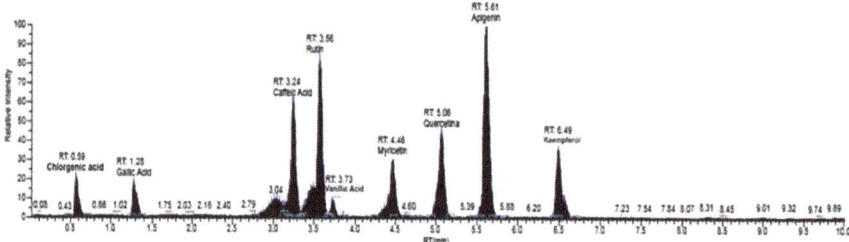

Figure 1. Representative LC/MS-MS of phenolic components in leaf methanolic extract of *M. styphelioides* leaves.

Previous studies regarding the phenolic components of *Melaleuca* plants have reported a rich composition of bioactive compounds, such as ellagitannins and flavonoids [15,16]. According to Al-Sayed et al. [15,16], gallic acid, kaempferol-3-O-α-L-rhamnopyranoside, pedunculagin, pterocarinin A, tellimagrandin I, casuarinin, cellimagrandin II, 1,2,3,6-tetra-O-galloyl-β-D-glucopyranose and 1,2,3,4,6-penta-O-galloyl-β-D-glucopyranose were the representative polyphenol compounds in *M. styphelioides* leaves. Al-Abd et al. [19] studied the phenolic composition of *Melaleuca cajuputi* leaves and indicate the presence of gingerol, caffeic acid, phenyl ester, aspidin, trans-2,3,4-trimethoxycinnamate, methyl orsellinic acid ester, ethyl ester, 5,6,3′-trimethoxyflavone, epigallocatechin 3-O-(4-hydroxybenzoate), polygonolide, and metyrosine. Compared to the aforementioned studies, our results show striking differences, presumably due to the species, origin, and analytical conditions [19]. Nevertheless, knowledge of *M. styphelioides* polyphenol compounds was extended by the identification of fourteen compounds (vanillic acid, gallic acid, caffeic acid, syringic acid, chlorogenic acid, ferulic acid, ellagic acid, apigenin, kaempferol, myricetin, naringenin, quercetin, luteolin, pinocembrin, hesperidin, and rutin) not previously reported in the *M. styphelioides* plant. From this finding, *M. styphelioides* appears as a source of bioactive compounds, justifying its pharmacological properties. Almost all of the identified phenolic compounds were well-known as antioxidants and anti-inflammatories.

The ability of *M. styphelioides* polyphenol extract to reduce the DPPH, ABTS radicals, and ferric reducing antioxidant power (FRAP) are shown in Table 3. DPPH and ABTS assays are frequently used as inexpensive, valid, and easy assays to evaluate the radical scavenging capacity of antioxidants. All extracts, excluding hexane extract, showed a concentration-dependent scavenging activity. The methanolic extract proved to be the highest DPPH and ABTS scavenger, with mean EC_{50} values of 22.13 and 21.39 µg/ml, respectively, followed by diethyl ether extract (73.24 ± 2.811 and 52.22 ± 1.40, respectively) and ethyl acetate (119.15 ± 1.66 and 75.84 ± 1.22, respectively). These results reflect the hydrogen-donating ability of the methanolic extracts from *M. styphelioides* leaves. On the other hand, the higher DPPH and ABTS scavenging activities of methanolic samples are most likely attributed to their higher total phenolic contents.

The ferric reducing power method involves the reduction of the Fe(III)-tripyridyltriazine (Fe(III)-TPTZ) complex to Fe(II)-tripyridyltriazine (Fe(II)-TPTZ) at a low pH by electron-donating antioxidants, resulting in the absorbance increase at λ = 593 nm. As presented in Table 3, the FRAP values of the methanolic extract (3.66 mM $FeSO_4$/g dw) was significantly ($p < 0.05$) higher than that ethyl acetate extract (0.85 mM $FeSO_4$/g dw). The non-polar extracts with diethyl ether and hexane did not present any reducing power activity. As displayed in Table 3, it is clearly observed that the methanolic extracts can be considered as effective scavengers of DPPH and ABTS radicals, as well as potent reducing agents.

From a mechanistic standpoint, the observed antioxidant activity reflects the ability of the test extracts to donate electrons or hydrogen atoms to inactivate radical species [20]. Such properties have been reported for numerous phenolic compounds, namely gallic acid, syringic acid, chlorogenic acid, caffeic acid, rutin, chlorogenic acid, luteolin glucoside, apigenin derivative, and quercetin [21,22].

In addition, our results confirm the potential therapeutic uses of *M. styphelioides* extract for inflammatory diseases and cancers [15,16]. Moreover, our data, supported by analytical LC/MS-MS, are in agreement with those of Al-Sayed et al. [15,16], although their data are limited to evaluation of the antiradical activity of methanolic extracts from *M. styphelioides* leaves by DPPH assay only.

Table 3. 2,2-diphenyl-1-picrylhydrazyl (DPPH) and 2,2′-azinobis-(3-ethylbenzothiazoline-6-sulphonate) (ABTS) scavenging activities, as well as the ferric reducing antioxidant power (FRAP) of *M. styphelioides* leaf extracts.

M. styphelioides Extracts	DPPH IC$_{50}$ µg/mL	ABTS IC$_{50}$ µg/mL	FRAP mM FeSO$_4$/g DE
E. MeOH	22.13 ± 2.17	21.39 ± 0.62	3.66 ± 0.014
E. EtOAc	119.15 ± 1.669	75.84 ± 1.22	0.85 ± 0.002
E. Et$_2$O	73.24 ± 2.811	52.22 ± 1.40	nd
E. Hexane	229.9 ± 5.8	201.35 ± 9.4	nd
Trolox IC$_{50}$	13.69 ± 0.04	64.37 ± 1.28	
BHT IC$_{50}$	19.33 ± 0.32		

FRAP FeSO$_4$·7H$_2$O equivalent mM per gram of dry extract (mM/g DE). nd: not determined.

Keratinocytes are the major players of the innate immune response in the skin. They promote inflammation via expression of several key bio-mediators. However, these cells can also adapt an anti-inflammatory behavior, in order to stop acute inflammation and retrieve a steady state through the secretion of immuno-modulating agents, such as ICAM-1, COX-2, iNOS, and NF-κB. Human keratinocytes have been reported to be useful cellular barrier models for both host–pathogen interaction studies and synthetic or natural compound anti-inflammatory screening. Here, we used NCTC 2544 cells to assess the effect of *M. styphelioides* phenolic extract on IFN-γ/H-induced inflammation.

The cell toxicity of polyphenol extracts was assessed by an MTT assay after 72 h treatment with different concentrations of *M. styphelioides* polyphenols. The MTT assay revealed that the MeOH extract had a lower toxicity compared to other extracts (data not shown). Moreover, the dose of 50 µg/mL of M. MeOH extract showed no toxicity and a high antioxidant property (data not shown). Thus, we chose to use 50 µg/mL concentration in NCTC 2544 cells treated with IFN-γ/H, in order to investigate the potential anti-inflammatory activity of the M. MeOH extract.

In order to examine the anti-inflammatory activity of M. MeOH extract in human keratinocytes, we measured in IFN-γ/H-treated NCTC 2544 cell proteins and mRNA gene expression levels of ICAM-1 and COX-2 by Western blot and RT-PCR, respectively. As shown in Figure 2A, IFN-γ/H treatment for 72 h significantly increased the level of mRNA expression of ICAM-1, whereas M. MeOH extract treatment (50 µg/mL) for 72 h inhibited the IFN-γ/H induced expression of ICAM-1. Compared to M. MeOH, indomethacin (10 µM) inhibited the mRNA expression of ICAM-1 to a higher degree. The Western blot confirmed the results obtained by RT-PCR (Figure 2B). In inflamed keratinocytes, the expression of ICAM-1 is always increased providing retention and activation of lymphocytes (CD4$^+$ and CD8$^+$) in the epidermis. Moreover, the regulation of ICAM-1 expression on keratinocytes using phenols is considered as a recent strategy for the treatment of skin inflammatory diseases [23].

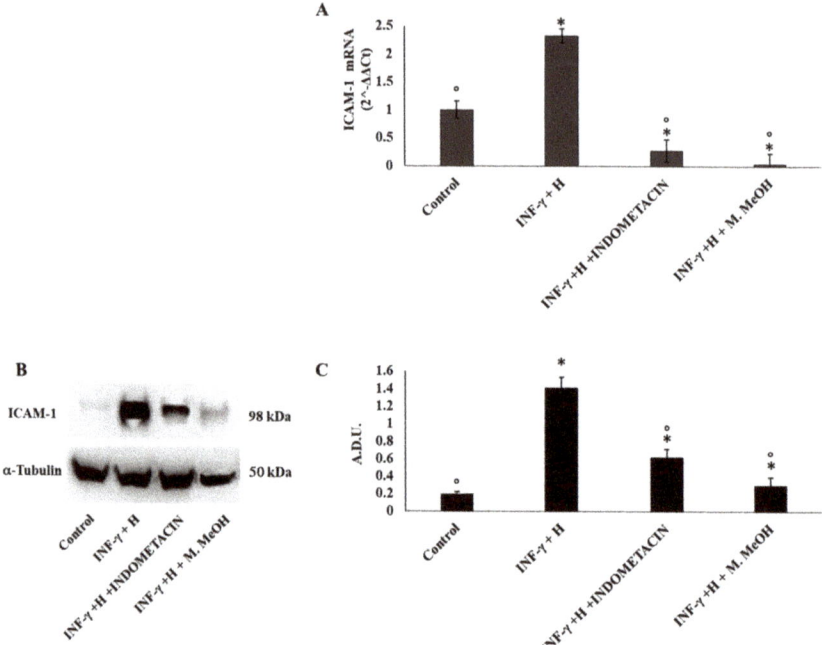

Figure 2. Inter-cellular adhesion molecule-1 (ICAM-1) mRNA expression (**A**) and protein production (**B**, and **C**). ICAM-1 mRNA expression was determined by RT-PCR (**A**), and ICAM-1 protein production was determined using Western blot (**B**: representative immunoblot; **C**: protein expression calculated as Arbitrary Densitometric Units; A.D.U.) in NCTC 2544 72 h after the addition of M. MeOH (50 μg/mL) with INF-γ + H. * Significantly different than control; ° significantly different from INF-γ/H-treated samples ($p < 0.05$).

Prostaglandins are also one of the major classes of mediators in the inflammatory response. They are generated from arachidonate by the action of cyclooxygenase isoenzymes (COX-1 and COX-2) [24]. In most tissues, COX-1 is constitutively expressed, whereas COX-2 is highly inducible by a variety of inflammatory and tumor-promoting stimuli, and is constitutively upregulated in skin carcinomas [25]. Our data demonstrates that COX-2 mRNA expressions were reduced significantly by the M. MeOH extract treatment in IFN-γ/H-induced inflammation, whereas its effect on protein levels was moderate compared to indomethacin (Figure 3).

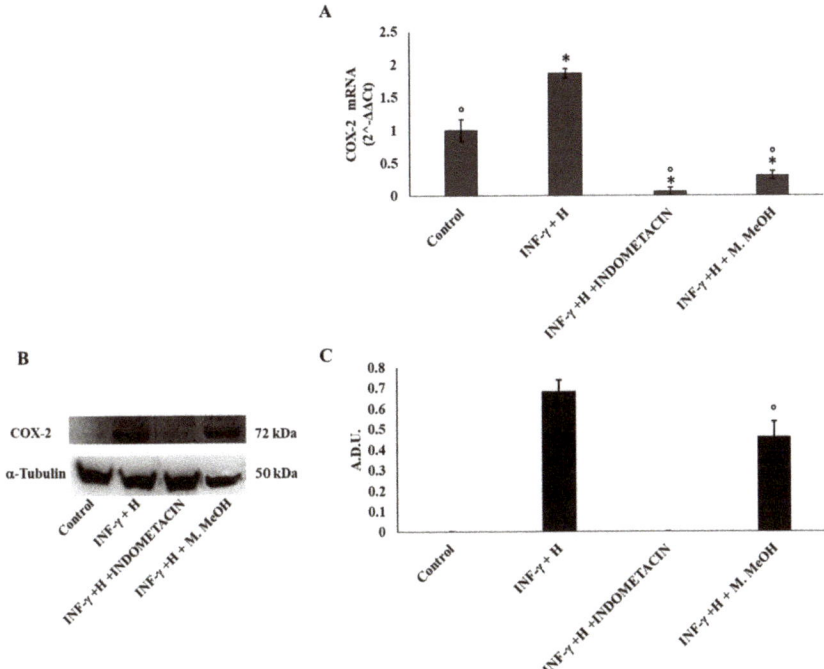

Figure 3. Cyclooxygenase-2 (COX-2) mRNA expression (**A**) and protein production (**B**, and **C**). COX-2 mRNA expression was determined by RT-PCR (**A**), and COX-2 protein production was determined using Western blot (**B**: representative immunoblot; **C**: protein expression calculated as Arbitrary Densitometric Units; A.D.U.) in the NCTC 2544 72 h after the addition of M. MeOH (50 μg/mL) with INF-γ + H. * Significantly different than control; ° significantly different than INF-γ/H-treated samples ($p < 0.05$).

The expression of the inducible nitric oxide synthase (iNOS) is one of the direct consequences of an inflammatory process. Several methods such as RT-PCR, immunocytochemistry, and Western blot, have been used to describe the iNOS expression in many chronic human inflammatory diseases. The iNOS mRNA level is mainly regulated at the transcriptional level but also by other post-transcriptional regulatory mechanisms. Phenolic acid and flavonoids have been reported to inhibit iNOS expression in several cell types, including endothelial cells, epithelial cells, and macrophages, yet in some cell types like chondrocytes [26]. Here, our data showed that the iNOS mRNA level was increased after cell stimulation with IFN-γ/H for 72 h, with respect to the non-treated control. A significant inhibition of iNOS mRNA expression was observed when the NCTC 2544 cells were co-treated with IFN-γ/H and 50 μg/mL of M. MeOH extract. Compared to M. MeOH extract, indomethacin weakly reduced iNOS mRNA levels (Figure 4B).

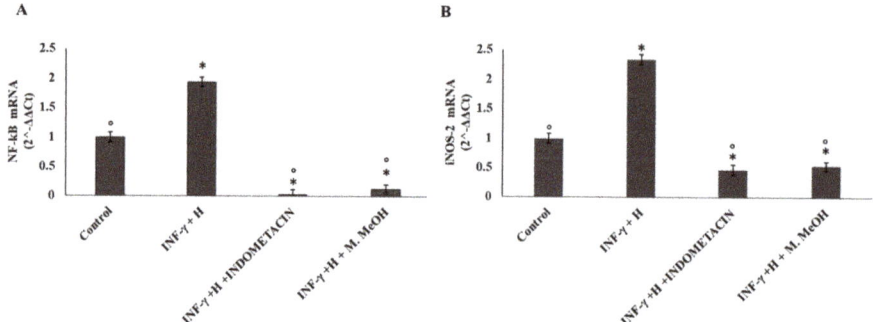

Figure 4. Nuclear factor kappa B (NF-κB) (**A**) and nitric oxide synthase (iNOS) mRNA expression (**B**). The mRNA expression was determined by RT-PCR in the NCTC 2544 for 72 h after the addition of M. MeOH (50µg/mL) with INF-γ/H. * Significantly different than control; ° significantly different from INF-γ/H-treated samples ($p < 0.05$).

In addition, we examined the influence of M. MeOH extract treatment on IFN-γ/H-induced NF-κB mRNA expression. As shown in Figure 4A, NF-κB mRNA expression was increased after stimulation with IFN-γ and H for 72 h. Significant inhibition of NF-κB mRNA expression was observed when the NCTC 2544 was co-treated with IFN-γ/H and 50 µg/mL of M. MeOH extract. Compared to M. MeOH extract, indomethacin weakly reduced the NF-κB mRNA levels. Several studies have provided that the keratinocytes, when exposed to IFN-γ and histamine, activate NF-κB, leading to the expression of inflammatory genes. More than 150 genes have been identified that are regulated by NF-κB activation. These genes include iNOS, ICAM-1, COX-2, and chemokines [27]. Therefore, NF-κB is primarily an inducer of inflammatory cytokines, and its inhibitors could be useful as anti-inflammatory agents. In the present study, our data showed that M. MeOH extracts inhibited the mRNA expression of NF-κB in IFN-γ/H-activated NCTC 2544 keratinocytes. In addition, iNOS, ICAM-1, and COX-2 mRNA expression were markedly decreased with the M. MeOH extract treatment. Compared with indomethacin, the effect of M. MeOH extract was either higher or similar, probably due to the different active compounds present in *M. styphelioides* extracts.

In summary, our work represents the first report of an anti-inflammatory effect of *M. styphelioides* extract in IFN-γ/H-stimulated human NCTC 2544 cells. M. MeOH extract inhibited the expression of pro-inflammatory mediators by inflamed NCTC 2544 cells. This inhibition was primarily mediated by the modulation of the major cellular effectors of inflammation, such as NF-κB, iNOS, iCAM-1, and COX-2.

Quercetin, the major flavonoid molecule found in M. MeOH extract (53.99%), has been reported to down-regulate the activation of NF-κB, ICAM-1 [23], iNOS [28], and COX-2 [29]. Gallic acid (13.5%) and ellagic acid (6.3%), two important compounds of M. MeOH extract, have been reported to produce a beneficial antioxidant effect by regulating several pathways. These include inhibition of COX-2 [30] and cytokines activated by the NF-κB pathway [31]. All compounds identified and quantified in M. MeOH extract, like vanillic acid [32,33], apigenin [34], syringic acid [35], kaempferol [36], and rutin [37] have also been shown to have a potential anti-inflammatory activity. Our data suggest that each molecule may have contributed to the anti-inflammatory effect of the M. MeOH extract. However, additional work is needed to better characterize the contribution of each compound, even though the phenolic compounds probably act synergistically to generate a more potent anti-inflammatory effect.

3. Conclusions

Previously, *M. styphelioides* methanolic extract has been investigated for anti-proliferative and anti-cancer effects [38], but to our knowledge, we have shown for the first time that *M. styphelioides* extract also has an anti-inflammatory property. The cytoprotective, antioxidant, and anti-inflammatory potential of *M. styphelioides* can be of great interest as an effective alternative for the treatment and prevention of inflammatory diseases.

4. Materials and Methods

4.1. Chemicals and Reagents

All chemicals and reagents were either analytical-reagent or HPLC grade. Ultrapure deionized water, with a resistivity of 18.2 MΩ cm, was obtained from a Milli-Q® Integral water purification system with a Q-pod purchased from Millipore (Bedford, MA, USA). Acetic acid, acetonitrile, and 2-propanol were purchased from VWR International S.r.l. (Milan, Italy); hydrochloric acid and sodium hydroxide were purchased from Carlo Erba (Milan, Italy). The methanol HPLC gradient grade was obtained from Merck (Darmstadt, Germany). Standard solutions of vanillic acid, trans-ferulic acid, syringic acid, myricetin, naringenin, pinocembrin, luteolin, and hesperidin were purchased from Extrasynthese (Genay Cedex, France). Chlorogenic acid was purchased from HWI Analytik GmbH (Rülzheim, Germany). Gallic acid, caffeic acid, apigenin, quercetin, kaempferol, catechin, epicatechin, indometacine, histamine, and anti-α-tubulin antibody were purchased from Sigma-Aldrich S.r.l. (Milan, Italy). The IFN-γ was obtained from Pepro Tech EC (London, England). The anti-ICAM-1 antibody (sc-51632) was purchased from Santa Cruz Biotechnology (Milan, Italy), anti-COX-2 (35-8200) from Thermo Fisher Scientific (Milan, Italy), and -α- tubulin (T6074) from Sigma-Aldrich (Milan, Italy).

4.2. Plant Material and Extracts Preparation

Leaves of *M. styphelioides* were collected from healthy trees in April 2016 from the botanical garden of the National Institute of Agricultural Research (INRAT, Tunis), Tunisia. Specimens were deposited at the Herbarium of the Department of Botany in the cited institute. Leaves were air-dried at room temperature (20 ± 2 °C) for one week, ground using a Retsch blender mill (Normandie-Labo, Normandy, France), sifted through a 0.5 mm mesh screen to obtain a uniform particle size, and subsequently assessed for their phenolic composition. Dried and ground leaves *M. styphelioides* were defatted three times by maceration with n-hexane for 48 h. The defatting process was used to extract lipophilic pigments and oil from lipid-containing samples and facilitate the polyphenol extract. The defatted samples were extracted with solvents of increasing polarity (diethyl ether, ethyl acetate, and methanol), sonicated for 30 min, and macerated for 24 h. The macerated organic extracts were filtered through Wattman filter paper, centrifuged for 10 min at 3000× g, and concentrated under reduced pressure in a Heidolph rotary evaporator. The obtained extracts were kept and stored at 4 °C until further analysis.

4.3. Phytochemical Analysis

4.3.1. The Phytochemical Screening

The phytochemical screening of total phenol, flavonoids, and tannins was determined according to the procedure reported by El Euch et al. [39].

4.3.2. LC/MS-MS Analysis

The mixture of polyphenols was determined according to previous work [40], using a Transcen II System with Multi-channel and Turbo Flow Technology (Dionex–Thermo Fisher Scientific) connected to a Q-Exactive Plus Hybrid Quadrupole-Orbitrap Mass Spectrometer (Thermo Fisher Scientific) equipped with HESI (heated electro spray ionization), used in negative polarity modes. The samples were extracted and purified using a Cyclone P column (50 mm × 0.5 m, 60 µm particle size, 60 Å pore size; Thermo Fisher Scientific). A Hypersil Gold (2.1 mm × 100 mm, 1.7 µm particle size) column was employed as the analytical separation column. The mobile phase consisted of eluent

A (30 mM ammonium acetate (pH 5)), eluent B (methanol), eluent C (water containing 0.5% formic acid), and eluent D (acetonitrile/acetone/2 propanol (4:3:3)). Mobile phases A and B were used to optimize the chromatographic resolution. Mobile phases B, C, and D were required for purification in turbo flow. Extracts and standards were dissolved in the mobile phase A (ratio 1:10). The injection volume was 5 µL, and elution was performed at the rate of 0.2 mL/min with a gradient program as follows: 0–2 min with 95% A and 5% B; 4 min with 60% A and 40% B; 6 min with 0% A and 100% B; 9 min with 0% A and 100% B; 11.5 min with 95% A and 5% B; 14 min with 95% A and 5% B; and 18 min with 95% A and 5% B. The acquisition time was 10 min. Eluted components were detected by MS, used in negative polarity modes, using the ion source parameters as follows: sheath gas flow rate of 35 (arbitrary units); aux gas flow rate of 10 (arbitrary units); spray voltage at 3.50 kV; capillary temperature at 300 °C; tube lens voltage of 55 V; heater temperature of 305 °C; scan mode at full scan; scan range (m/z) at 100–700; microscans at 1 m/z; a positive resolution of 70,000; an FT automatic gain control (AGC) target of 3×10^6; a maximum IT of 100 ms; a negative resolution of 35,000; an automatic gain control (AGC) target of 1×10^6; and a maximum IT of 100 ms. The chromatographic parameters were as follows: column temperature at 30 °C and sample temperature at 6 °C. The auto-sampler sample holder temperature was maintained at 7 °C. The data analyses were performed using a Thermo Scientific XCalibur (Thermo Fisher Scientific) version 4.0 software and Qual and Quant Browser, and the concentration of the compounds were calculated using calibration curves; the results are expressed as calibration curves and as µg/kg of dry weight (DW). This method was validated according to the norm EN ISO/IEC 17025:2005. The limits of detection and quantification (LoDs and LoQs) were determined by the 3σ and 10σ approach [41,42]. The calibration curve was constructed with six standard additions (0.05, 0.1, 0.2, 0.5, 1, and 2 mg/L), and was checked using the r^2 value. The linearity range was considered to be acceptable when r^2 was greater than 0.99 in the peak areas versus the concentration. A pool of 15 blank samples spiked with final concentrations of 5 µg/kg for all elements were analysed. The results were between 10 µg/kg for the LOD and 20 µg/kg for the LOQ for single analytes. The limit of repeatability and recovery has been evaluated, with the spike samples at three different concentration levels (0.1, 0.5, and 1 mg/L). The results were satisfactory for the limit of repeatability (metrological approach), less than the double value of the expanded uncertainty. The recovery was between 71% and 119%. The validation allowed us to identify the uncertainty contributions in order to calculate the expanded uncertainty [43]. The results show that the expanded uncertainty is less than 22% in all the analyzed levels.

4.4. Anti-Oxidant Activity Determination

The DPPH and ABTS radical scavenging activity was assessed by the method described by El Euch et al. [39]. The antioxidant effect was expressed as IC_{50}, which is the amount of extract required to scavenge the initial DPPH or ABTS radical by 50%, and is expressed as per mg of the sample [39]. The reducing power (FRAP) was determined by the method described in [44]. The results were expressed as $FeSO_4 \cdot 7H_2O$-equivalent mM per gram of dry extract (mM/g DE), using a calibration curve.

4.5. Anti-Inflammatory Activity Evaluation

4.5.1. Cell Culture and Treatment

The NCTC 2544 keratinocyte cell line was obtained from Interlab Cell Line Collection (Genoa, Italy). Cells were grown in Minimum Essential Medium (Sigma-Aldrich, Milan, Italy), containing 10% foetal bovine serum (FBS), 100 µg/mL streptomycin, and 100 U/mL penicillin. The cells were then incubated at 37 °C in a humidified, 95% air 5% CO_2 atmosphere. The culture medium was changed every 2–3 days. For experiments, cells were trypsinized, counted, and plated in six- or 96-well plates. Cells were either treated or not treated with 200 U mL^{-1} of IFN-γ and 10^{-4} M of H, in the presence or absence of different concentrations of *M. styphelioides* polyphenol extracts (5, 10, 25,

50, and 75 µg/mL). Commercially available indomethacin was used as a positive control at 10 µM. After 72 h, each sample was tested for the experiments described below.

4.5.2. Cell Viability

The phenol extracts' cytotoxicity was determined by an MTT assay, as previously described [45].

4.5.3. RNA Extraction and RT-PCR

As previously described [46], the total mRNA was isolated and prepared from the control and treated NCTC 2544 cells using the 1 mL Qiazol Reagent (Qiagen, Milan, Italy), 0.2 mL chloroform, and 0.5 mL isopropanol. An RNA pellet was washed with 75% ethanol, air-dried, and re-suspended in RNAse-free water. Reverse transcription was carried using the QuantiTect Reverse Transcription Kit (Qiagen), according to the manufacturer's protocol. Synthesis of cDNA was performed using 40-cycle PCR in a Rotor-gene Q real-time analyzer (Corbett, Qiagen). The amplification of ICAM-1, iNOS, COX-2, NF-κB, and GAPDH was performed using specific primers listed in Table 4. Each PCR reaction contained one Rotor-Gene SYBR Green PCR Master Mix, template cDNA ($\leq$100 ng/reaction), primers (1 µM), and RNase-free water, with a final reaction volume of 25 µL. RT-PCR was carried out according to the following program: initial activation step at 95 °C for 10 min, denaturation at 95 °C for 10 s, annealing at 60 °C for 30 s, extension at 72 °C for 30 s (40 cycles), and final extension at 72 °C for 10 min. RT-PCR was followed by melting curve analysis to confirm PCR specificity. Each reaction was repeated three times, and the threshold cycle average was used for data analysis by Rotor-gene Q software. The identification was carried out using electrophoresis in a 2% agarose gel in 0.045 M Tris–borate/1 mM EDTA (TBE) buffer. The target gene expression was normalized to GAPDH using the $2^{-\Delta\Delta Ct}$ method.

Table 4. Primers used in RT-PCR analysis.

Primers	Forward (5′→3′)	Reverse (5′→3′)
ICAM-1	GGCCGGCCAGCTTATACAC	TAGACACTTGAGCTCGGGCA
iNOS	GTTCTCAAGGCACAGGTCTC	GCAGGTCACTTATGTCACTTATC
NF-κB	ATGGCTTCTATGAGGCTGAG	GTTGTTGTTGGTCTGGATGC
COX-2	ATCATTCACCAGGCAAATTGC	GGCTTCAGCATAAAGCGTTTG
GAPDH	TCAACAGCGACACCCAC	GGGTCTCTCTCTTCCTCTTGTG

4.5.4. Western Blot

The ICAM-1 and COX-2 protein expression levels were determined by Western blot analysis according to standard procedures [47]. The NCTC 2544 cells were stimulated with IFN-γ/H and treated with 50 µg/mL of *M. styphelioides* polyphenol extract for 72 h at 37 °C. Western blot was performed following the method reported [24]. Briefly, equal protein amounts were resolved by 4–12% Novex Bis–Tris gel electrophoresis (NuPAGE, InVitrogen, Milan, Italy), and transferred into nitrocellulose membranes (InVitrogen) in a wet system. The membrane was blocked in Tris-buffered saline containing 0.01% Tween-20 (TBST) and 5% non-fat dry milk for 1 h at room temperature. The membrane was incubated overnight with specific primary antibodies at 4 °C. Mouse monoclonal anti-ICAM-1 (1:200) (1H4: sc-51632; Santa Cruz Biothechnology), anti-COX-2 (1:300) (35-8200; Thermo Fisher Scientific), and anti-α-tubulin antibodies (Sigma-Aldrich) were used. Blots were later washed three times with PBS, followed by incubation in a HRP-conjugated secondary antibody for 1 h at room temperature. Specific proteins bands were detected using enhanced chemiluminescent solution (Pierce, Fisher Scientific) and visualized by a Uvitec Alliance LD9 gel imaging system (Uvitec, Cambridge, UK). Bands were measured densitometrically, and their relative density was calculated based on the density of α-tubulin bands in each sample. Values were expressed as arbitrary densitometric units (A.D.U.) corresponding to signal intensity.

4.6. Statistical Analysis

The experiments were repeated independently at least three times in triplicate, and the mean ± SEM for each value was calculated. One-way statistical analyses of the results (Student's *t*-test for paired and analysis of variance (ANOVA) for unpaired data) were used. All statistical analyses were performed using the statistical software package SYSTAT, version 11 (Systat Inc., Evanston, IL, USA). A value of $p < 0.05$ was considered statistically significant.

Author Contributions: Conceptualization, F.A. and V.C. (Venera Cardile); Funding acquisition, V.C. (Venera Cardile); Investigation, F.A., R.A. and A.C.E.G.; Methodology, F.A., G.M.L.D. and A.C.E.G.; Supervision, V.C. (Venera Cardile); Validation, R.A., M.A. and V.C. (Venera Cardile); Visualization, V.C. (Vittorio Calabrese); Writing—original draft, F.A.; Writing—review & editing, V.C. (Venera Cardile).

Acknowledgments: The authors wish to thank Filippo Drago (Director of Department of Biomedical and Biotechnological Science, University of Catania, Catania, Italy) for his hospitality and scientific encouragement.

Conflicts of Interest: The authors declare that there is no conflict of interests regarding the publication of this paper.

Abbreviations

(H)	Histamine
(IFN-γ)	interferon gamma
(NF-κB)	nuclear factor kappa B
(ICAM-1)	inter-cellular adhesion molecule-1
(iNOS)	nitric oxide synthase
(COX-2)	cyclooxygenase-2
(DPPH)	2,2-diphenyl-1-picrylhydrazyl
(ABTS)	2,2′-azinobis-(3-ethylbenzothiazoline-6-sulphonate)
(E. hex)	hexane extract
(E. Et2O)	diethyl ether extract
(E. EtOAc)	ethyl acetate extract
(E. MeOH)	methanol extract
(M. MeOH)	*Melaleuca styphelioides* methanolic extract

References

1. Medzhitov, R. Origin and physiological roles of inflammation. *Nature* **2008**, *454*, 428–435. [CrossRef] [PubMed]
2. Köllisch, G.; Kalali, B.N.; Voelcker, V.; Wallich, R.; Behrendt, H.; Ring, J.; Bauer, S.; Jakob, T.; Mempel, M.; Ollert, M. Various members of the Toll-like receptor family contribute to the innate immune response of human epidermal keratinocytes. *Immunology* **2005**, *114*, 531–541. [CrossRef] [PubMed]
3. McInturff, J.E.; Modlin, R.L.; Kim, J. The role of toll-like receptors in the pathogenesis and treatment of dermatological disease. *J. Investig. Dermatol.* **2005**, *125*, 1–8. [CrossRef] [PubMed]
4. Pastore, S.; Mascia, F.; Mariani, V.; Girolomoni, G. The epidermal growth factor receptor system in skin repair and inflammation. *J. Investig. Dermatol.* **2008**, *128*, 1365–1374. [CrossRef] [PubMed]
5. Mascia, F.; Mariani, V.; Girolomoni, G.; Pastore, S. Blockade of the EGF receptor induces a deranged chemokine expression in keratinocytes leading to enhanced skin inflammation. *Am. J. Pathol.* **2003**, *163*, 303–312. [CrossRef]
6. Kerkvliet, N.I. AHR-mediated immunomodulation: The role of altered gene transcription. *Biochem. Pharmacol.* **2009**, *77*, 746–760. [CrossRef] [PubMed]
7. Pastore, S.; Mascia, F.; Mariotti, F.; Dattilo, C.; Mariani, V.; Girolomoni, G. ERK1/2 regulates epidermal chemokine expression and skin inflammation. *J. Immunol.* **2005**, *174*, 5047–5056. [CrossRef] [PubMed]
8. Essafi-Benkhadir, K.; Refai, A.; Riahi, I.; Fattouch, S.; Karoui, H.; Essafi, M. Quince (Cydonia oblonga Miller) peel polyphenols modulate LPS-induced inflammation in human THP-1-derived macrophages through NF-κB, p38MAPK and Akt inhibition. *Biochem. Biophys. Res. Commun.* **2012**, *418*, 180–185. [CrossRef] [PubMed]

9. Korkina, L.G. Phenylpropanoids as naturally occurring antioxidants: From plant defense to human health. *Cell. Mol. Biol.* **2017**, *53*, 15–25. [CrossRef]
10. Yang, H.J.; Kim, M.J.; Kang, S.; Moon, N.R.; Kim, D.S.; Lee, N.R.; Kim, K.S.; Park, S. Topical treatments of Saussurea costus root and *Thuja orientalis* L. synergistically alleviate atopic dermatitis-like skin lesions by inhibiting protease-activated receptor-2 and NF-κB signaling in HaCaT cells and Nc/Nga mice. *J. Ethnopharmacol.* **2017**, *199*, 97–105. [CrossRef] [PubMed]
11. Kim, D.O.; Lee, K.W.; Lee, H.J.; Lee, C.Y. Vitamin C equivalent antioxidant capacity (VCEAC) of phenolic phytochemicals. *J. Agric. Food Chem.* **2002**, *50*, 3713–3717. [CrossRef] [PubMed]
12. Korkina, L.G.; Pastore, S.; De Luca, C.; Kostyuk, V.A. Metabolism of plant polyphenols in the skin: Beneficial versus deleterious effects. *Curr. Drug Metab.* **2008**, *9*, 710–729. [CrossRef] [PubMed]
13. Wu, N.; Kong, Y.; Fu, Y.; Zu, Y.; Yang, Z.; Yang, M.; Peng, X.; Efferth, T. In vitro antioxidant properties, DNA damage protective activity, and xanthine oxidase inhibitory effect of cajaninstilbene acid, a stilbene compound derived from pigeon pea [*Cajanus cajan* (L.) Millsp.] leaves. *J. Agric. Food Chem.* **2011**, *59*, 437–443. [CrossRef] [PubMed]
14. Sharifi-Rad, J.; Salehi, B.; Varoni, E.M.; Sharopov, F.; Yousaf, Z.; Ayatollahi, S.A.; Kobarfard, F.; Sharifi-Rad, M.; Afdjei, M.H.; Sharifi-Rad, M.; et al. Plants of the Melaleuca Genus as Antimicrobial Agents: From Farm to Pharmacy. *Phyther. Res.* **2017**, *31*, 1475–1494. [CrossRef] [PubMed]
15. Al-Sayed, E.; El-Lakkany, N.M.; Seif El-Din, S.H.; Sabra, A.N.A.; Hammam, O.A. Hepatoprotective and antioxidant activity of *Melaleuca styphelioides* on carbon tetrachloride-induced hepatotoxicity in mice. *Pharm. Biol.* **2014**, *52*, 1581–1590. [CrossRef] [PubMed]
16. Al-Sayed, E.; Esmat, A. Hepatoprotective and antioxidant effect of ellagitannins and galloyl esters isolated from Melaleuca styphelioides on carbon tetrachloride-induced hepatotoxicity in HepG2 cells. *Pharm. Biol.* **2016**, *54*, 1727–1735. [CrossRef] [PubMed]
17. Farag, R.S.; Shalaby, A.S.; El-Baroty, G.A.; Ibrahim, N.A.; Ali, M.A.; Hassan, E.M. Chemical and Biological Evaluation of the Essential Oils of Different Melaleuca Species. *Phyther. Res.* **2004**, *18*, 30–35. [CrossRef] [PubMed]
18. Amri, I.; Mancini, E.; de Martino, L.; Marandino, A.; Lamia, H.; Mohsen, H.; Bassem, J.; Scognamiglio, M.; Reverchon, E.; de Feo, V. Chemical composition and biological activities of the essential oils from three Melaleuca species grown in Tunisia. *Int. J. Mol. Sci.* **2012**, *13*, 16580–16591. [CrossRef] [PubMed]
19. Al-Abd, N.M.; Mohamed Nor, Z.; Mansor, M.; Azhar, F.; Hasan, M.S.; Kassim, M. Antioxidant, antibacterial activity, and phytochemical characterization of *Melaleuca cajuputi* extract. *BMC Complement. Altern. Med.* **2015**, *15*. [CrossRef] [PubMed]
20. Yuan, Y.V.; Bone, D.E.; Carrington, M.F. Antioxidant activity of dulse (*Palmaria palmata*) extract evaluated in vitro. *Food Chem.* **2005**, *91*, 485–494. [CrossRef]
21. Albouchi, F.; Hassen, I.; Casabianca, H.; Hosni, K. Phytochemicals, antioxidant, antimicrobial and phytotoxic activities of *Ailanthus altissima* (Mill.) Swingle leaves. *S. Afr. J. Bot.* **2013**, *87*, 164–174. [CrossRef]
22. Lesjak, M.; Beara, I.; Simin, N.; Pintać, D.; Majkić, T.; Bekvalac, K.; Orčić, D.; Mimica-Dukić, N. Antioxidant and anti-inflammatory activities of quercetin and its derivatives. *J. Funct. Foods* **2018**, *40*, 68–75. [CrossRef]
23. Graziano, A.C.E.; Cardile, V.; Crascì, L.; Caggia, S.; Dugo, P.; Bonina, F.; Panico, A. Protective effects of an extract from *Citrus bergamia* against inflammatory injury in interferon-gamma and histamine exposed human keratinocytes. *Life Sci.* **2012**, *90*, 968–974. [CrossRef] [PubMed]
24. Smith, W.L.; DeWitt, D.L.; Garavito, R.M. Cyclooxygenases: Structural, Cellular, and Molecular Biology. *Annu. Rev. Biochem.* **2000**, *69*, 145–182. [CrossRef] [PubMed]
25. Müller-Decker, K.; Scholz, K.; Neufang, G.; Marks, F.; Fürstenberger, G. Localization of prostaglandin-H synthase-1 and -2 in mouse skin: Implications for cutaneous function. *Exp. Cell Res.* **1998**, *242*, 84–91. [CrossRef] [PubMed]
26. Suschek, C.; Schnorr, O.; Kolb-Bachofen, V. The Role of iNOS in Chronic Inflammatory Processes In Vivo: Is it Damage-Promoting, Protective, or Active at all? *Curr. Mol. Med.* **2004**, *4*, 763–775. [CrossRef] [PubMed]
27. Kumar, A.; Takada, Y.; Boriek, A.M.; Aggarwal, B.B. Nuclear factor-kappaB: Its role in health and disease. *J. Mol. Med.* **2004**, *82*, 434–448. [CrossRef] [PubMed]
28. García-Mediavilla, V.; Crespo, I.; Collado, P.S.; Esteller, A.; Sánchez-Campos, S.; Tuñón, M.J.; González-Gallego, J. The anti-inflammatory flavones quercetin and kaempferol cause inhibition of inducible

nitric oxide synthase, cyclooxygenase-2 and reactive C-protein, and down-regulation of the nuclear factor kappaB pathway in Chang Liver cells. *Eur. J. Pharmacol.* **2007**, *557*, 221–229. [CrossRef] [PubMed]
29. Xiao, X.; Shi, D.; Liu, L.; Wang, J.; Xie, X.; Kang, T.; Deng, W. Quercetin suppresses cyclooxygenase-2 expression and angiogenesis through inactivation of P300 signaling. *PLoS ONE* **2011**, *6*, e22934. [CrossRef] [PubMed]
30. Chatterjee, A.; Chatterjee, S.; Das, S.; Saha, A.; Chattopadhyay, S.; Bandyopadhyay, S.K. Ellagic acid facilitates indomethacin-induced gastric ulcer healing via COX-2 up-regulation. *Acta Biochim. Biophys. Sin.* **2012**, *44*, 565–576. [CrossRef] [PubMed]
31. Ahad, A.; Ganai, A.A.; Mujeeb, M.; Siddiqui, W.A. Ellagic acid, an NF-κB inhibitor, ameliorates renal function in experimental diabetic nephropathy. *Chem. Biol. Interact.* **2014**, *219*, 64–75. [CrossRef] [PubMed]
32. Kim, H.J.; Chen, F.; Wu, C.; Wang, X.; Chung, H.Y.; Jin, Z. Evaluation of antioxidant activity of Australian tea tree (*Melaleuca alternifolia*) oil and its components. *J. Agric. Food Chem.* **2004**, *52*, 2849–2854. [CrossRef] [PubMed]
33. Kim, M.C.; Kim, S.J.; Kim, D.S.; Jeon, Y.D.; Park, S.J.; Lee, H.S.; Um, J.Y.; Hong, S.H. Vanillic acid inhibits inflammatory mediators by suppressing NF-κB in lipopolysaccharide-stimulated mouse peritoneal macrophages. *Immunopharmacol. Immunotoxicol.* **2011**, *33*, 525–532. [CrossRef] [PubMed]
34. Wang, J.; Liu, Y.T.; Xiao, L.; Zhu, L.; Wang, Q.; Yan, T. Anti-Inflammatory Effects of Apigenin in Lipopolysaccharide-Induced Inflammatory in Acute Lung Injury by Suppressing COX-2 and NF-kB Pathway. *Inflammation* **2014**, *37*, 2085–2090. [CrossRef] [PubMed]
35. Ham, J.R.; Lee, H.-I.; Choi, R.-Y.; Sim, M.-O.; Seo, K.-I.; Lee, M.-K. Anti-steatotic and anti-inflammatory roles of syringic acid in high-fat diet-induced obese mice. *Food Funct.* **2016**, *7*, 689–697. [CrossRef] [PubMed]
36. Kadioglu, O.; Nass, J.; Saeed, M.E.M.; Schuler, B.; Efferth, T. Kaempferol is an anti-inflammatory compound with activity towards NF-κB pathway proteins. *Anticancer Res.* **2015**, *35*, 2645–2650. [PubMed]
37. Yoo, H.; Ku, S.K.; Baek, Y.D.; Bae, J.S. Anti-inflammatory effects of rutin on HMGB1-induced inflammatory responses in vitro and in vivo. *Inflamm. Res.* **2014**, *63*, 197–206. [CrossRef] [PubMed]
38. Ganesan, D.; Al-Sayed, E.; Albert, A.; Paul, E.; Singab, A.N.B.; Govindan Sadasivam, S.; Saso, L. Antioxidant activity of phenolic compounds from extracts of *Eucalyptus globulus* and *Melaleuca styphelioides* and their protective role on D-glucose-induced hyperglycemic stress and oxalate stress in NRK-49Fcells. *Nat. Prod. Res.* **2018**, *32*, 1274–1280. [CrossRef] [PubMed]
39. El Euch, S.K.; Bouajila, J.; Bouzouita, N. Chemical composition, biological and cytotoxic activities of Cistus salviifolius flower buds and leaves extracts. *Ind. Crops Prod.* **2015**, *76*, 1100–1105. [CrossRef]
40. López-Gutiérrez, N.; del MAguilera-Luiz, M.; Romero-González, R.; Vidal, J.L.M.; Garrido Frenich, A. Fast analysis of polyphenols in royal jelly products using automated TurboFlow™-liquid chromatography-Orbitrap high resolution mass spectrometry. *J. Chromatogr. B Anal. Technol. Biomed. Life Sci.* **2014**, *973*, 17–28. [CrossRef] [PubMed]
41. International Standard Organization. *ISO/IEC 17025 General Requirements for the Competence of Testing and Calibration Laboratories*; ISO: Geneva, Switzerland, 2005; pp. 1–36.
42. Eurachem, a Laboratory Guide to Method Validation and Related Topics. 2014. Available online: http://www.eurachem.org/images/stories/Guides/pdf/valid.pdf (accessed on 25 July 2017).
43. Lo Dico, G.M.; Cammilleri, G.; Macaluso, A. Simultaneous Determination of As, Cu, Cr, Se, Sn, Cd, Sb and Pb Levels in Infant Formulas by ICP-MS after Microwave-Assisted Digestion: Method Validation. *J. Environ. Anal. Toxicol.* **2015**, *5*, 1–5. [CrossRef]
44. Le Man, H.; Behera, S.K.; Park, H.S. Optimization of operational parameters for ethanol production from korean food waste leachate. *Int. J. Environ. Sci. Technol.* **2010**, *7*, 157–164. [CrossRef]
45. Graziano, A.C.E.; Parenti, R.; Avola, R.; Cardile, V. Krabbe disease: Involvement of connexin43 in the apoptotic effects of sphingolipid psychosine on mouse oligodendrocyte precursors. *Apoptosis* **2016**, *21*, 25–35. [CrossRef] [PubMed]

46. Avola, R.; Graziano, A.C.E.; Pannuzzo, G.; Albouchi, F.; Cardile, V. New insights on Parkinson's disease from differentiation of SH-SY5Y into dopaminergic neurons: An involvement of aquaporin4 and 9. *Mol. Cell. Neurosci.* **2018**, *88*, 212–221. [CrossRef] [PubMed]
47. Avola, R.; Graziano, A.C.E.; Pannuzzo, G.; Cardile, V. Human Mesenchymal Stem Cells from Adipose Tissue Differentiated into Neuronal or Glial Phenotype Express Different Aquaporins. *Mol. Neurobiol.* **2017**, *54*, 8308–8320. [CrossRef] [PubMed]

Sample Availability: Samples of the compounds are available from the authors.

© 2018 by the authors. Licensee MDPI, Basel, Switzerland. This article is an open access article distributed under the terms and conditions of the Creative Commons Attribution (CC BY) license (http://creativecommons.org/licenses/by/4.0/).

Review

Aza- and Azo-Stilbenes: Bio-Isosteric Analogs of Resveratrol

Gérard Lizard [1], Norbert Latruffe [1] and Dominique Vervandier-Fasseur [2,*]

[1] Team Bio-PeroxIL, Biochemistry of the Peroxisome, Inflammation and Lipid Metabolism (EA7270), University of Bourgogne Franche-Comté, Inserm, 21000 Dijon, France; gerard.lizard@u-bourgogne.fr (G.L.); norbert.latruffe@u-bourgogne.fr (N.L.)
[2] Team OCS, Institute of Molecular Chemistry of University of Burgundy (ICMUB UMR CNRS 6302), University of Bourgogne Franche-Comté, 21000 Dijon, France
* Correspondence: dominique.vervandier-fasseur@u-bourgogne.fr; Tel.: +33-3-8039-9036

Academic Editor: Corrado Tringali
Received: 19 December 2019; Accepted: 23 January 2020; Published: 30 January 2020

Abstract: Several series of natural polyphenols are described for their biological and therapeutic potential. Natural stilbenoid polyphenols, such as trans-resveratrol, pterostilbene and piceatannol are well-known for their numerous biological activities. However, their moderate bio-availabilities, especially for trans-resveratrol, prompted numerous research groups to investigate innovative and relevant synthetic resveratrol derivatives. This review is focused on isosteric resveratrol analogs aza-stilbenes and azo-stilbenes in which the C=C bond between both aromatic rings was replaced with C=N or N=N bonds, respectively. In each series, synthetic ways will be displayed, and structural sights will be highlighted and compared with those of resveratrol. The biological activities of some of these molecules will be presented as well as their potential therapeutic applications. In some cases, structure-activity relationships will be discussed.

Keywords: trans-resveratrol; aza-stilbene; azo-stilbene; bio-isosterism; structure-activity relationship

1. Introduction

Among natural polyphenolic compounds, polyphenolic stilbenoids hold an important place. They are widespread in a large number of plants. The leader of this series is a phytoalexin, trans-resveratrol (RSV) or 3,4′,5-trihydroxystilbene (1) (Figure 1) discovered in 1940 in Japan by M. Takaoka from *Veratrum grandiflorum* where its name comes from: VERATRum/resVERATRol [1].

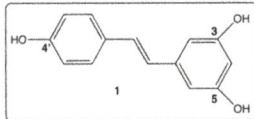

Figure 1. Structure of *trans*-Resveratrol (**1**).

This natural polyphenol is found in numerous species, for instance, in roots of Asiatic plant *Polygonum cuspidatum* [2], in several edible plants [3] especially in fruit including grapes where RSV is in the form of phytoalexin [4,5], and subsequently in red wine [6]. RSV is widely studied since the 90s. Because of its numerous biological activities, such as anti-oxidant [7], antitumoral [8], antiviral [9], and anti-inflammatory activities [10] and more recently due to its differentiating properties [11,12]. In addition, trans-resveratrol is a neuroprotective agent [13], and acts against platelet aggregation [14]. RSV is a sirtuin-activating compound (STAC) which may increase lifespan in metazoans (*Caenorhabditis elegans*, *Drosophila melanogaster*, mice) by a mechanism related with a caloric restriction [15–17].

Subsequently, this polyphenol is effective in treating metabolic disorders [18]. It is now recognized that a diet rich in these health-beneficial molecules allows a good level of health to be maintained [19].

However, despite its therapeutic potential, so far RSV could not be used as such in clinical trials because its quick metabolism and its weak bio-availability due to its low water solubility [20,21]. In order to enhance the bio-availability of RSV and subsequently, its therapeutic potency while keeping the hydroxylated stilbene scaffold, numerous research groups have synthesized an infinite number of RSV analogs and evaluated them for various biological activities. Several ways to modify RSV may be considered, for instance, transformation of phenolic functions in ester or ether functions [22–24], substitution with various groups on the phenyl rings [25–28] or by replacing a phenyl ring with an aromatic heterocycle [29,30] or with an organometallic cycle [27]. The concept of bio-isosterism may be applied in the case of RSV analogs [31]. Indeed, the C=C bond may be seen as a bridge between both phenyl rings, allowing an electronic delocalization on the whole molecule. This electronic feature plays a primordial role especially in anti-oxidant activity of the polyphenolic molecule. Thus, some RSV analogs have been designed by replacing the C=C bond with isosteric C=N or N=N bonds or with an aromatic ring [32–35]. Another way to overcome the problems related to the low water solubility of RSV is to load it into nanoparticles or liposomes [36–39]. Several reviews have focused on several series of RSV analogs especially natural RSV analogs [24], RSV analogs displaying pharmacological activities [40,41], multi-targeted drug RSV analogs [42] and RSV analogs with various substituents on both phenyl rings [28].

The aim of this review is to specifically focus on both aza-stilbenes (AZA-ST) and azo-stilbenes (AZO-ST) whose C=C bond is replaced by a C=N bond or N=N bond, respectively (Figure 2). Each series will be examined with respect to synthetic ways as well as structural sights and biological activities. In some cases, a relationship between the latter will be highlighted.

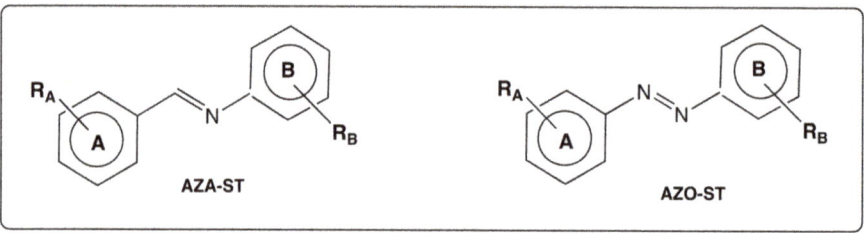

Figure 2. Structure of aza-stilbenes (AZA-ST) and azo-stilbenes (AZO-ST).

2. Structural and Synthetic Sights of AZA-ST and AZO-ST Compared with trans-RSV

2.1. Isosteric Features of C=N and N=N Bonds

Numerous RSV derivatives have been designed by keeping the original stilbenoid skeleton and changing the nature and/or the number of substituents on aromatic rings in order to enhance the bio-availability of the molecule, to better target the receptors, or to carry pharmacophore in the case of multi-targeted ligand derivatives [28]. These modifications usually require multi-steps syntheses which are often easy to carry out. The bio-isosterism concept is a discerning tool in drug design and may be used in the case of RSV analogs. According to Grimm's hydride displacement law [31], nitrogen atom and CH group having the same number of valence electrons are isosters. Thus, the replacement of one or both CH in the double bond of stilbene by one or two nitrogen atoms is an easy synthetic way to obtain bio-isosters of RSV. In such isosters, the number of valence electrons does not change with respect to the parent molecule electronic environment of the stilbene scaffold and subsequently, the resulting RSV analogs may have biological activities similar to, or even superior to RSV.

2.2. Synthetic Pathways for Obtaining AZA-ST and AZO-ST

Among the numerous synthetic ways to get RSV derivatives bearing a C=C bond, the key-step is always the formation of this linkage which, can be formed by different chemical methods including Perkin [43], Wittig [25], Horner-Wittig-Emmons [44], Heck [25] and Suzuki [45] reactions from starting aromatic aldehydes or aryl bromide (Figure 3). These methods are summarized in the figure below (Figure 3); some of them require protection steps of phenolic functions.

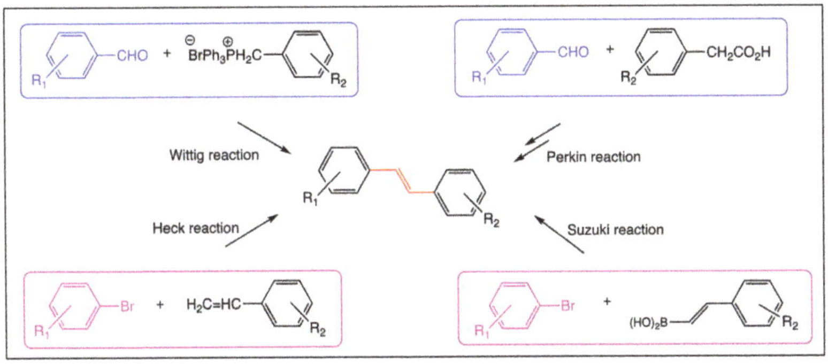

Figure 3. Principal synthetic methods for obtaining stilbene derivatives: Wittig method [25], Perkin method [43], Heck method [25] and Suzuki method [45].

Aza-stilbenes (AZA-ST) are commonly obtained by one-step reactions between aromatic aldehydes and primary aromatic amines. These one-step reactions may be carried out in refluxing ethanol [46], in refluxing toluene in a Dean–Stark apparatus [47] or in water at 25 °C during 2 h [48] or 3 days [33] (Scheme 1). The reagents used to carry out these straightforward condensation reactions are mostly commercially available allowing to obtain large series of aza-stilbenes, most of them being substituted with hydroxyl groups.

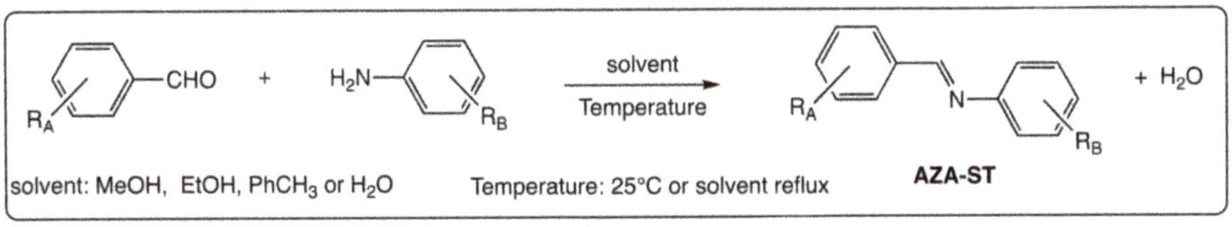

Scheme 1. Synthetic methods for obtaining aza-stilbenes (AZA-ST).

In contrast, the synthesis of azo-stilbenes (AZO-ST) requires several steps: preparation of a diazonium salt from an aromatic primary amine followed by diazo-coupling reaction between this salt and an aromatic compound [32,49] (Scheme 2). As in the case of AZA-ST, most AZO-ST are substituted with hydroxyl groups. Thus, the library of AZO-ST is large because of the commercial availability of the reagents.

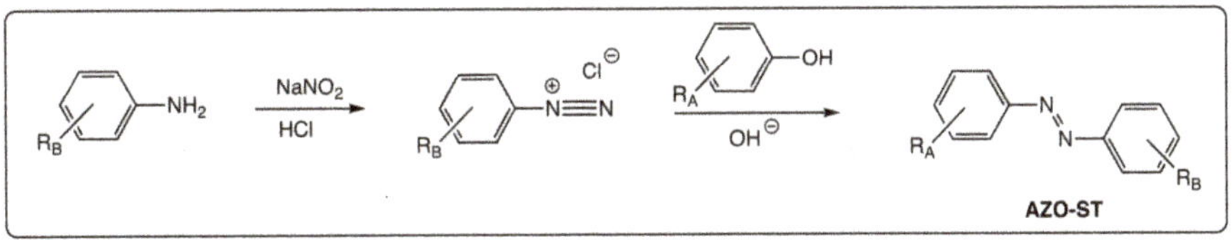

Scheme 2. Synthetic methods for obtaining Azo-stilbenes (AZO-ST).

However, reported works about aza-stilbenes are more numerous than those concerning azo-stilbenes, probably because of their asymmetrical double bond which may afford more structural, chemical and biological specifications.

2.3. Symmetry or Dissymmetry of Double Bonds

The N=N bond in AZO-ST is symmetric as in the case of RSV and does not provide an electronic influence on one or the other of the aromatic rings, especially on their substituents. In contrast, the presence of imino bond C=N in AZA-ST induces a dissymmetry inside the stilbene core. According to the number and positions of hydroxyl groups on both aromatic rings A and B (Figure 2), their position with respect to the nitrogen atom in the linkage, mono or poly hydroxy AZA-ST may give rise to various biological studies. Indeed, the lone electronic pair of the nitrogen atom can play a role in the stabilization of a phenoxyl radical or may allow an intramolecular hydrogen bond. In addition, the imino bond is polarized and the carbon atom may be attacked by a nucleophilic agent, such as a thiol group of a cysteine residue. These features that do not appear in the case of RSV impute to aza-stilbenes specific behaviors in biological environment (Scheme 3).

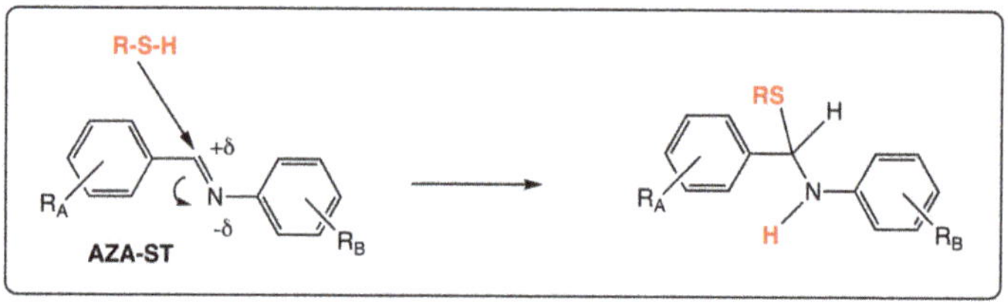

Scheme 3. Example of the nucleophilic attack of a thiol on an imino bond in AZA-ST.

3. Biological Activities of Aza-Stilbenes

3.1. Aza-Stilbenes Bearing A Hydroxyl Group in Ortho Position of Cycles A or/and B

Anti-oxidant activity is widespread in polyphenols and especially in the case of resveratrol. At first, the comparison between RSV and its imino analogs has been focused on anti-oxidant and radical scavenging activities by different research groups. The high anti-oxidant activities of AZA-ST **2** towards 2,2-diphenyl-1-picrylhydrazyl (DPPH) radical and galvinoxyl (GO) radical have been explained by the acidity of the phenolic proton due to an intramolecular hydrogen bond between this phenolic proton and the nitrogen atom in the imine group (Figure 4) [48]. In addition, the ^{1}H-NMR spectrum of **2** showed high chemical shifts for such phenolic protons (δ = 12.25 to 14.19 ppm) that are typical of acidic protons.

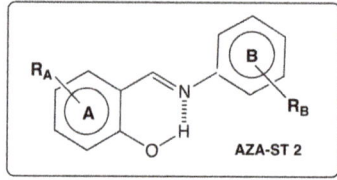

Figure 4. AZA-ST **2** bearing OH group in ortho position in ring A [48].

The authors have suggested the following mechanism in three steps [48]. The spontaneous release of the phenolic proton leading to the formation of a phenolate anion may trigger the anti-oxidant mechanism in AZA-ST **2** (ArOH → ArO$^-$ + H$^+$). Indeed, the phenolate anion loses an electron in favor of a free radical R$^\bullet$ (ArO$^-$ + R$^\bullet$ → ArO$^\bullet$ + R$^-$) especially since the phenoxyl radical is stabilized on the aromatic ring and anion R$^-$ is neutralized by proton provided in the first step (R$^-$ + H$^+$ → R – H).

Radical scavenging activities (RSA) of compounds of AZA-ST series **3** (Figure 5) against DPPH radical have been studied [50,51]. Imino RSV analogs **3** bearing a hydroxyl group in ortho position of cycle B have shown a better activity than the parent molecule [51]. The lone electronic pair of nitrogen atom would overlap with the phenoxyl radical at *ortho* position of the cycle B, initiated by the scavenging of DPPH radical. On the other hand, the presence of a catechol group on cycle A in AZA-ST **3a** increased the anti-oxidant activity because the resulting phenoxyl radical may be stabilized by resonance and formed subsequently a *o*-quinone [52]. Indeed, compound **3a** was the most effective anti-oxidant agent of this series against DPPH radical and provided an IC$_{50}$ value (expressed in µM) 6-fold lower than RSV. In addition, compounds in the AZA-ST series **3** were evaluated for their ability to quench singlet oxygen $^1O_2$ by using EPR spin-trapping technique. All of them appeared to be better quenchers than RSV, especially compound **3b**, whose IC50 value (expressed in µM) was 15-fold lower compared to that of the parent molecule [51]: the IC50 of compound **3b** determined with the EPR spin-trapping technique was (0.99 ± 0.06) µM whereas the IC50 value of RSV was (16.94 ± 0.73) µM.

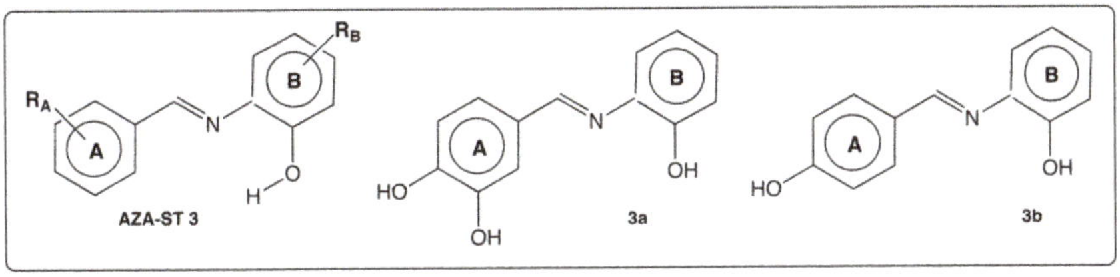

Figure 5. Structure of compounds of AZA-ST series **3** and AZA-ST **3a** and **3b** bearing OH in *ortho* position in ring **B** [51].

In addition to anti-oxidant features of imino RSV analogs, Zhang's group has shown a correlation between radical scavenging activities of different aza-stilbenes and their abilities to chelate transition metal ions such as Cu^{2+} and Fe^{3+}, especially in the case of AZA-ST **4a** (Figure 6) [53]. The authors have suggested that the "N=C=C-OH" sequence (shown in red in Figure 6) was a "metal ion-binding motif".

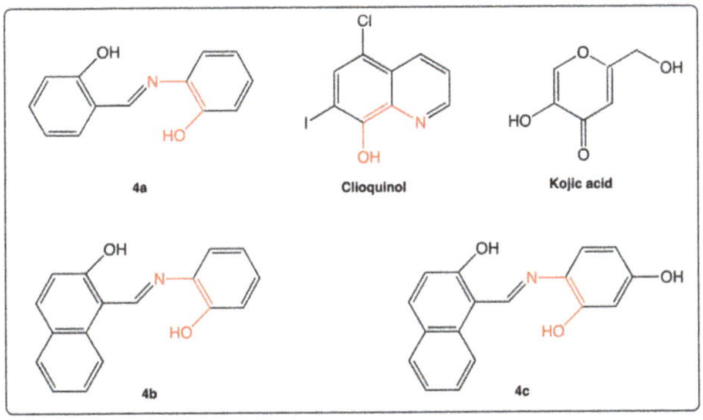

Figure 6. Structure of clioquinol, kojic acid and AZA-ST **4a–c**.

Free radicals and transition metal ions including Cu^{2+}, Fe^{2+} and Fe^{3+} catalyze oxidative damages as does the Fenton reaction ($Fe^{2+} + H_2O_2 \rightarrow Fe^{3+} + OH^- + HO^\bullet$) in some age-related diseases by initiating decomposition reactions of hydrogen peroxide (H_2O_2) with metal ions to generate the hydroxyl radical ($HO^\bullet$), which is a powerful pro-oxidant. Therefore, the AZA-ST **4b–c** (Figure 6) were designed by conjugation of RSV and clioquinol, both compounds being different agents against Alzheimer's disease. Indeed, in vitro, RSV is known for its inhibition of the aggregation of amyloid-β (Aβ) [54,55], the main component of amyloid plaques in Alzheimer's diseases; clioquinol, bearing a metal ion-binding motif, is fit to slow down the neurological decline in early stage clinical trials [56]. To combine both features and strengthen the activity against Alzheimer's disease, ionophoric polyphenols **4b** and **4c**, bearing a hydroxyl group in *ortho* position of both aromatic cycles, were synthesized [57]. The *ortho* position of hydroxyl group on ring B was essential to keep the same ability than clioquinol to chelate Cu^{2+} for an efficient activity. In contrast, the involvement of the lone electronic pair of nitrogen in the metal complexation prevents intramolecular hydrogen bonding with hydroxyl group in position *ortho* of cycle A as in the case of AZA-ST **2** [48]. However, this phenolic group is crucial for scavenging free radicals produced during the interaction between abnormal amyloid-β (Aβ) and Cu^{2+} [58].

AZA-ST **4a** has been evaluated for its ability to inhibit tyrosinase. Indeed, tyrosinase is a copper-containing protein; it is implied in the melanin biosynthesis in melanocytes and subsequently in hyperpigmentation of the skin. Chelators of copper ions, such as kojic acid (Figure 6) are good candidates to inhibit tyrosinase action [59]. Given its feature to bind Cu^{2+} ion [53], AZA-ST **4a** (Figure 6) was evaluated for its ability to inhibit tyrosinase by Lima's group [60]. Compound **4a** turned out to provide a lower tyrosinase inhibitory activity than kojic acid; however, compound **4a** showed a better depigmenting activity than RSV.

The compounds in the compounds of the AZA-ST series **5** (Figure 7) were tested for their antioxidant activities and compared to RSV [61]. They were more effective DPPH radical scavengers than RSV. In addition, compounds **5a** and **5e** turned to provide anti-inflammatory properties.

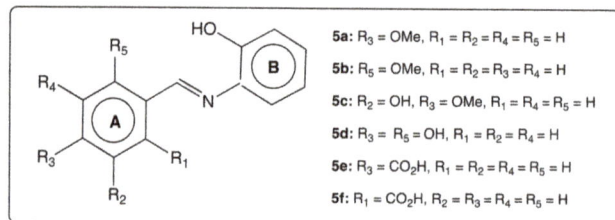

Figure 7. Structure of the compounds of the AZA-ST series **5** (AZA-ST **5**) [61].

Apart from the fact that the lone pair of nitrogen atom may play crucial roles in anti-oxidant activities and metal chelation, polarization of the imino bond providing an electrophilic character to the carbon atom is essential in reactions with nucleophilic agents, such as a thiol function of the cysteine residues in proteins. Among such proteins, Keap-1 (Kelch-like ECH-associated protein 1) is a major repressor of Nrf2 (Nuclear factor erythroid-2-related factor 2), a transcription factor regulating the expression of several genes encoding for enzymes involved in the control of RedOx homeostasis. Indeed, Keap-1 may form a complex with Nrf2, that prevents the transcription factor to bind to antioxidant response elements (ARE) in the nucleus. Li's group has shown the ability of AZA-ST **3a** and **6** (Figure 8) to activate Nrf2 and proposed the following mechanism: Keap-1 is kept close to AZA-ST **3a** or **6** by interactions between the protein and hydroxyl (or methoxyl for **6**) groups in *para* and *meta* positions of cycle A and *ortho* position of cycle B [33]. These interactions give a conformation to the aza-stilbene as the thiol group of cysteine residue may attack the carbon atom of the imino linkage to form a covalent bond and cause the release of Nrf2 (Figure 8) [33]. In this case, hydroxyl groups of both aromatic cycles are involved either upon interactions between the lone pairs of oxygen atoms and Keap-1 either by intramolecular hydrogen bonds with this protein.

Other compounds of a new AZA-ST series **7** (Figure 9) bearing a hydroxyl group in the *ortho* position of cycle B were synthesized and have shown promising anti-leishmanicidal and antituberculosis activities [62]. The authors suggest that activity of compounds **7a–7f** (which appear to be moderate) depends on the electronic density of the substituent in *para* position of cycle A. In contrast, the role of the hydroxyl group in cycle B was not mentioned.

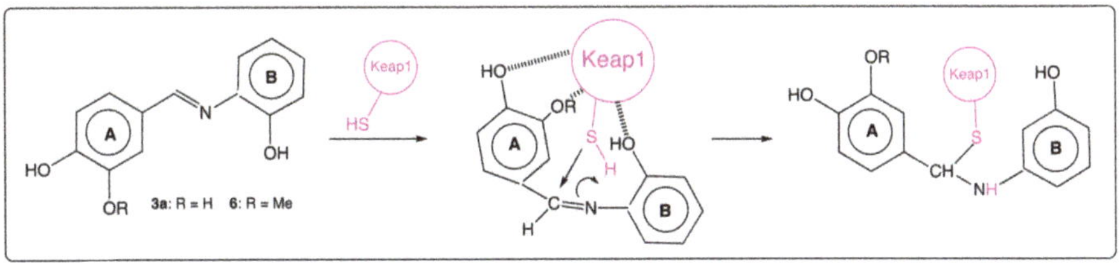

Figure 8. Suggested mechanism in the adduct of Keap1 protein on AZA-ST **3a** and **6**. (Inspired from [33]).

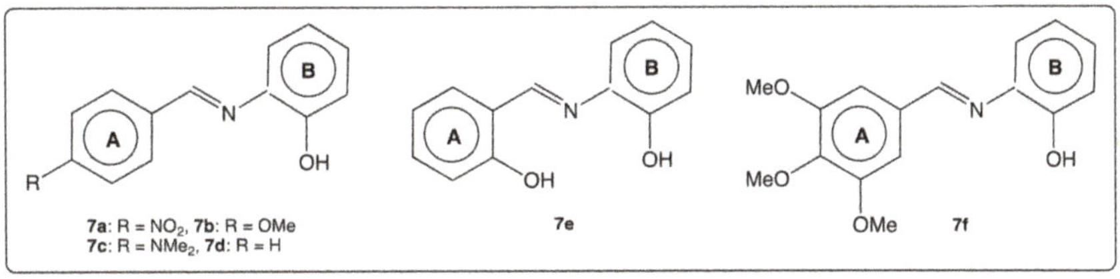

Figure 9. Structure of AZA-ST **7a–f** [62].

3.2. Aza-Stilbenes Bearing A Catechol Group on Cycle A

Murias's group has reported several studies on polyhydroxylated stilbenes bearing catechol or pyragollol groups on one or both aromatic cycles. They have highlighted the crucial roles of such groups in anti-oxidant activities and subsequently anti-tumoral activities of such polyhydroxylated stilbenes compared to that of RSV, which bears a resorcinol group (Figure 10) [63]. Indeed, as it was demonstrated by Wright [52], a phenoxyl radical formed from a catechol group may rise to a stabilized *ortho*-quinone, that it is not achievable with a resorcinol group. Thus, polyhydroxylated stilbenes bearing a catechol or a pyrogallol group provide better antioxidant and antitumoral activities than RSV.

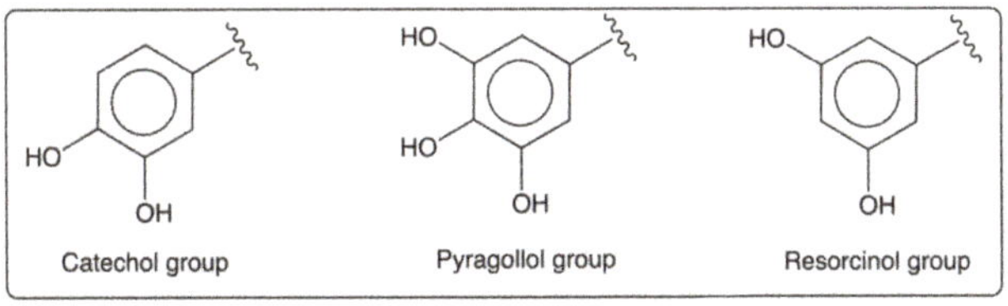

Figure 10. Catechol, pyrogallol and resorcinol groups [55].

In the case of catecholic aza-stilbenes, we have mentioned the above example of aza-stilbene **3a** (Figure 5) providing efficient anti-oxidant activities [51]. However, this feature was attributed to both the catechol group on cycle A and the phenol group in *ortho* position of cycle B. In the following examples, it will be shown that only catechol group on cycle A may be responsible for biological activities. Catechol compounds AZA-ST **8a–b** and phenol compounds AZA-ST **9a–e** (Figure 11) were synthesized and evaluated for their ability to scavenge GO radical [64]. The authors have shown that GO radical scavenging reaction rates of **8a–b** were higher than those of RSV and **9a–e**. Lu's group also reported a weak DPPH radical scavenging activity for AZA-ST **9a–e** [51]. In addition, **8a–b** provided better antiproliferative activity against human hepatoma HepG2 cells than RSV and **9a–e** with IC_{50} values 14-fold and 11.7-fold lower than RSV IC_{50} values, respectively. Regarding catecholic AZA-ST **8a–b**, the results highlight a correlation between anti-oxidant and anti-proliferative activities as well as the crucial role of the catechol group in anti-oxidant activities of such RSV analogs.

Bhat's group has reported the ability of AZA-ST **8a** and **8c** (Figure 11) to inhibit the growth of human breast cancer cell lines (MDA-MB-231, which is estrogen receptors (ER) negative and expresses mutated p53, and T47D, which is ERα positive) [46,65,66]. A docking study has allowed to show that (thanks to catechol groups) Van der Waals bonds might be promoted between the amino acid residues of protein in receptor ERα cavity and AZA-ST **8a** and **8c**. They would lead to a better stabilization of the imino RSV analogs into the protein pocket than RSV itself [46]. In subsequent studies, the same authors have highlighted that AZA-ST **8a** and **8c** might act on both estrogen receptors by inhibiting ERα expression and promoting ERβ expression [65,66].

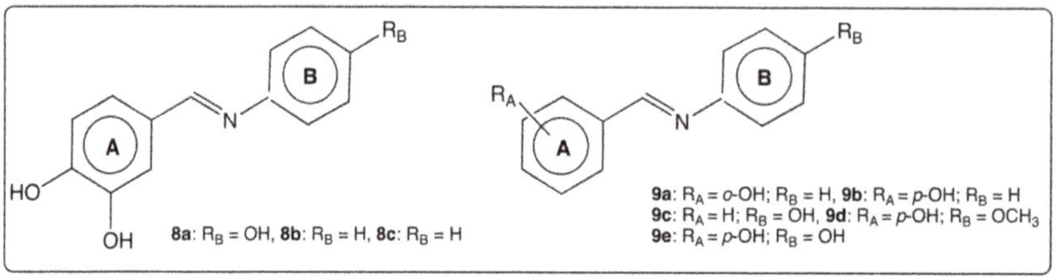

Figure 11. Structure of AZA-ST **8a–c** and **9a–e** [64].

4. Biological Activities of Azo-Stilbenes AZO-ST

Two series of azo-stilbenes AZO-ST **10** and AZO-ST **11** (Figure 12) were synthesized and evaluated in vitro for their antifungal activities against seven phytopathogenic fungi [67]. AZO-ST **11**, bearing a hydroxyl group in the *ortho* position of aromatic cycle A were less efficient than hymexazol the commercially agricultural fungicide (Figure 12). In contrast, AZO-ST **10** was efficient towards such fungi; some of them were more efficient than hymexazol. The substitution by a hydroxyl group in the *para* position of the aromatic cycle A and additionally by a methyl group in the *ortho* position of cycle A, seemed to be crucial structural parameters for AZO-ST **10** to get promising antifungal properties.

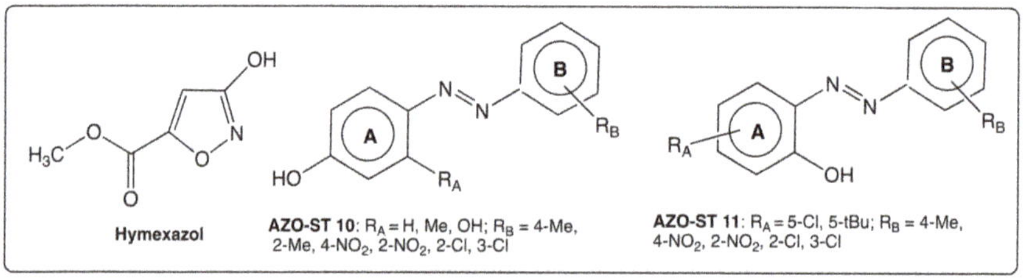

Figure 12. Structure of hymexazol and structure of compounds of the AZO-ST series **10** and AZO-ST series **11** [67].

Azo-stilbenes AZO-ST **12** (Figure 13) substituted with hydroxyl or methoxyl groups were synthesized and tested in vitro as potent tyrosinase inhibitors [49]. Azo-resveratrol (**12a**) and azo-pterostilbene (**12b**) with the *para* hydroxy group on cycle A had the best inhibition activities against mushroom tyrosinase in this series similar to those of RSV itself. In Figure 13, it is highlighted that the slightest modification of a group and of its position on both aromatic rings may have a great impact on the ability to inhibit tyrosinase. It can also be observed that even the induction of the *para* hydroxyl group is central, the role of the *ortho* hydroxyl group on cycle A is moderate and may be easily disturbed by a methoxyl group lying nearby.

The same research group has refined the structure of the compounds of the AZO-ST series **12** taking into account both important effects of the *para* hydroxyl group and the moderate effect of the *ortho* hydroxyl group in cycle A in the inhibitory activity on mushroom tyrosinase [32]. Four new compounds in the AZA-ST series **13a–d** (Figure 14) were synthesized, evaluated *in vitro* as potent mushroom tyrosinase inhibitors and compared with kojic acid (Figure 6) and RSV. Overall, compounds **13a–d** are more active than both reference compounds and IC$_{50}$ values are 17.85, 49.08 and 59.80 µM for **13b**, kojic acid and RSV, respectively. Tyrosinase exists widely in bacteria, fungi, plants, insects, vertebrates and invertebrates and is the rate limiting enzyme in the biosynthesis of melanin pigments responsible for colors in hair, skin and eyes. In such series, there are no more methoxyl groups whereas the number of hydroxyl group drops while two compounds bear a tosyl-oxy group. Thus, molecular structures being more stripped towards those of AZA-ST **12**, it is easier to see the influence of *para* and/or *ortho* hydroxyl groups. The weak percentage of tyrosinase inhibition obtained with **13a** shows that the combination of *para* and *ortho* hydroxyl groups is required to get potent tyrosinase inhibitors and subsequently, promising therapeutic agents to treat skin diseases (hyperpigmentation, lentigo, vitiligo and skin cancers) [68].

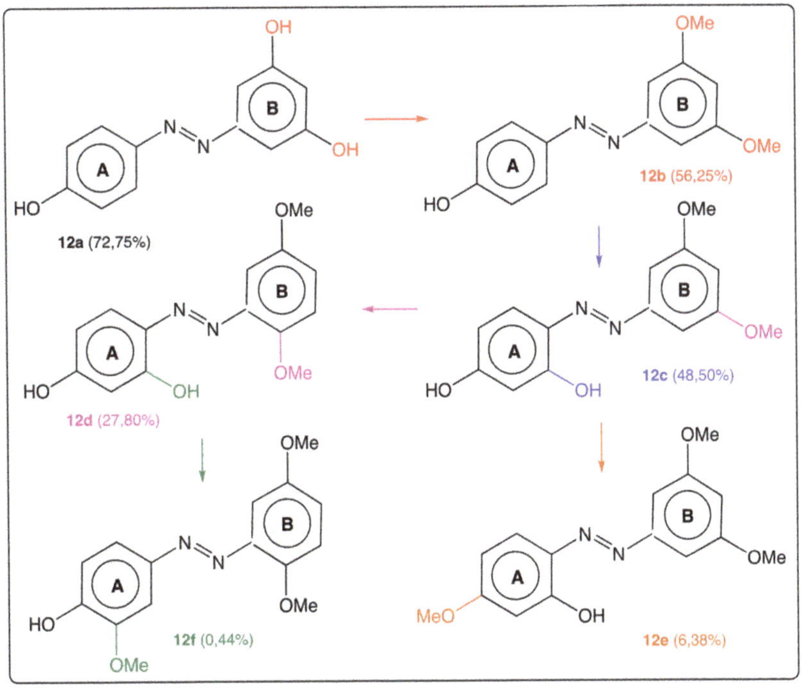

Figure 13. Structure of AZO-ST **12a–f** (and percentage of tyrosinase inhibition) [49].

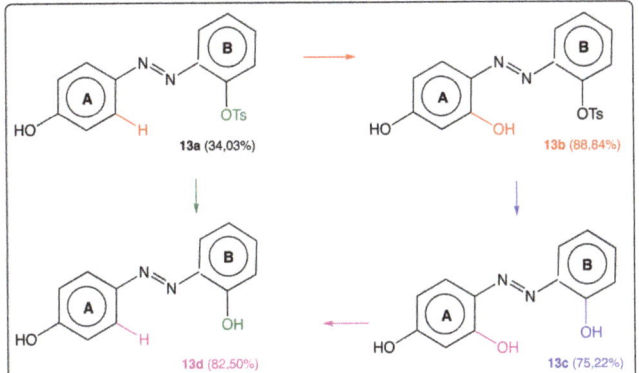

Figure 14. Structure of AZO-ST **13a–d** (and percentage of tyrosinase inhibition) [32].

5. Conclusions

AZA-ST and AZO-ST are isosteric RSV derivatives in which the lone electronic pair(s) of nitrogen atom(s) bring different electronic effects in the molecular structure compared to that of RSV. In the case of aza-stilbenes, the dissymmetric imino linkage induces a polarization of this bond and subsequent nucleophilic attacks [33]. The replacement of one or both carbon atoms of the double bond between the mono- or poly-hydroxylated aromatic rings provides attractive biological properties to AZA-ST and AZO-ST such as antioxidant activities [51,64]. Some of them may be promising candidates to treat diseases whose current treatments are either non-existent or weakly efficient especially breast cancer with ERα negative [46,65,66], leishmaniasis and tuberculosis [62] and skin diseases [68]. Finally, the compounds of AZO-ST series **10** and **11** also show greater anti-fungal activities than hymexazol, a commercially available agricultural fungicide [67]. Some diazo compounds are photoactivatable and may interact with proteins and nucleic acids. Hence, such features should be taken into account to avoid interferences with biochemical tests used to define their biological activities [69]. Given the easy syntheses and biological potential of AZA-ST and AZO-ST, the development of such isosteric RSV derivatives may be a perspective of interest to highlight promising therapeutic and fungicide agents.

Author Contributions: D.V.-F., G.L. and N.L. conceptualized this review; the state of the art and bibliographic work were conducted by D.V.-F.; writing—review and editing, D.V.-F., G.L. and N.L.; supervision, D.V.-F. All authors have read and agreed to the published version of the manuscript.

Funding: This work was jointly supported by the COST Action CA16112 NutRedOx, the CNRS, the Université de Bourgogne (CoMUE BFC), the Conseil Régional through the "Plan d'Action Régionale pour l'Innovation" (PARI) and the European Union through the PO FEDER-FSE Bourgogne 2014/2020.

Acknowledgments: Richard Decréau is acknowledged for English corrections.

Conflicts of Interest: The authors declare no conflict of interest.

References

1. Takaoka, M. Phenolic substances of white hellebore (*Veratrum grandiflorum* Loes. fil.). *J. Faculty Sci.* **1940**, *3*, 1–16. [CrossRef]
2. Arichi, H.; Kimura, Y.; Okuda, H.; Baba, M.; Kozowa, K.; Arichi, S. Effects of stilbene compounds of the roots of *Polygonum cuspidatum* Sieb. et Zucc. on lipid metabolism. *Chem. Pharm. Bull.* **1982**, *30*, 1766–1770. [CrossRef] [PubMed]
3. Burns, J.; Yokota, T.; Ashihara, H.; Lean, M.E.J.; Crozier, A. Plant foods and herbal sources of resveratrol. *J. Agric. Food Chem.* **2002**, *50*, 3337–3340. [CrossRef] [PubMed]
4. Adrian, M.; Jeandet, P. Effects of resveratrol on the ultrastructure of *Botrytis cinerea* conidia and biological significance in plant/pathogen interactions. *Fitoterapia* **2012**, *83*, 1345–1350. [CrossRef]
5. Pawlus, A.D.; Sahli, R.; Bisson, J.; Rivière, C.; Delaunay, J.C.; Richard, T.; Gomes, E.; Bordenave, L.; Waffo-Teguo, P.; Mérillon, J.M. Stilbenoid profiless of canes from *Vitis* and *Muscadinia* species. *J. Agric. Food Chem.* **2013**, *61*, 501–515. [CrossRef]
6. Boutegrabet, L.; Fekete, A.; Hertkorn, N.; Papastamoulis, Y.; Waffo-Téguo, P.; Mérillon, J.M.; Jeandet, P.; Gougeon, R.D.; Schmitt-Koplin, P. Determination of stilbene derivatives in Burgundy red wines by ultra-high pressure liquid chromatography. *Anal. Bioanal. Chem.* **2011**, *401*, 1517–1525. [CrossRef]
7. Gülçin, I. Antioxidant properties of resveratrol: A structure-activity insight. *Innov. Food Sci. Emerg. Technol.* **2010**, *11*, 210–218. [CrossRef]
8. Khan, O.S.; Bhat, A.A.; Krishnankutty, R.; Mohammad, R.M.; Uddin, S. Therapeutic potential of resveratrol in lymphoid malignancies. *Nutr. Cancer* **2016**, *68*, 365–373. [CrossRef]
9. Yiu, C.Y.; Chen, S.Y.; Chang, L.K.; Chiu, Y.F.; Lin, T.P. Inhibitory effects of resveratrol on the Epstein-Barr virus lytic cycle. *Molecules* **2010**, *15*, 7115–7124. [CrossRef]
10. Tili, E.; Michaille, J.J.; Adair, B.; Alder, H.; Limagne, E.; Taccioli, C.; Ferracin, M.; Delmas, D.; Latruffe, N.; Croce, C.M. Resveratrol decreases the levels of miR-155 by upregulating miR-663, a microRNA targeting *JunB* and *JunD*. *Carcinogenesis* **2010**, *31*, 1561–1566. [CrossRef]

11. Kaminski, J.; Lançon, A.; Aires, V.; Limagne, E.; Tili, E.; Michaille, J.J.; Latruffe, N. Resveratrol initiates differentiation of mouse skeletal muscle-derived C2C12 myoblasts. *Biochem. Pharmacol.* **2012**, *84*, 1251–1259. [CrossRef] [PubMed]
12. Namsi, A.; Nury, T.; Hamdouni, H.; Yammine, A.; Vejux, A.; Vervandier-Fasseur, D.; Latruffe, N.; Masmoudi-Kouki, O.; Lizard, G. Induction of neuronal differentiation of murine N2a cells by two polyphenols present in the mediterranean diet mimicking neurotrophins activities: Resveratrol and apigenin. *Diseases* **2018**, *6*, 67. [CrossRef] [PubMed]
13. Singh, N.; Agrawal, M.; Doré, S. Neuroprotective properties and mechanisms of resveratrol in in vitro and in vivo experimental cerebral stroke models. *ACS Chem. Neurosci.* **2013**, *4*, 1151–1162. [CrossRef] [PubMed]
14. Stef, G.; Csiszar, A.; Lerea, K.; Ungvari, Z.; Veress, G. Resveratrol inhibits aggregation of platelets from high-risk cardiac patients with aspirin resistance. *J. Cardiovasc. Pharmacol.* **2006**, *48*, 1–5. [CrossRef]
15. Wood, J.G.; Rogina, B.; Lavu, S.; Howitz, K.; Helfand, S.L.; Tatar, M.; Sinclair, D. Sirtuin activators mimic caloric restriction and delay ageing in metazoans. *Nature* **2004**, *430*, 686–689. [CrossRef]
16. Markus, M.A.; Morris, B.J. Resveratrol in prevention and treatment of common clinical conditions of aging. *Clin. Interv. Aging* **2008**, *3*, 331–339.
17. Westphal, C.H.; Dipp, M.A.; Guarente, L. A therapeutic role for situins in diseases of aging? *Trends Biochem. Res.* **2007**, *32*, 555–560. [CrossRef]
18. Bonkowski, M.S.; Sinclair, D.A. Slowing ageing by design: The rise of NAD^+ and sirtuin-activating compounds. *Nat. Rev. Mol. Cell Biol.* **2016**, *17*, 679–690. [CrossRef]
19. Salvatore Benito, A.; Valero Zanuy, M.Á.; Alarza Cano, M.; Ruiz Alonso, A.; Alda Bravo, I.; Rogero Blanco, E.; Maíz Jiménez, M.; León Sanz, M. Adherence to Mediterranean diet: A comparison of patients with head and neck cancer and healthy population. *Endocrinol. Diabetes Nutr.* **2019**, *66*, 417–424. [CrossRef]
20. Sun, X.; Peng, B.; Yan, W. Measurement and correlation of solubility of *trans*-resveratrol in 11 solvents at T = (278.2, 282.2, 298.2, 308.2 and 318.2) K. *J. Chem. Thermodyn.* **2008**, *40*, 735–738. [CrossRef]
21. Delmas, D.; Aires, V.; Limagne, E.; Dutartre, P.; Mazué, F.; Ghiringhelli, F.; Latruffe, N. Transport, stability and biological activity of resveratrol. *Ann. N. Y. Acad. Sci.* **2011**, *1215*, 48–59. [CrossRef] [PubMed]
22. Cardile, V.; Chillemi, R.; Lombardo, L.; Sciuto, S.; Spatafora, C.; Tringali, C. Antiproliferative activity of methylated analogues of *E*- and *Z*-resveratrol. *Z. Naturforsch. C* **2007**, *62*, 189–195. [CrossRef] [PubMed]
23. Liu, Q.; Kim, C.T.; Jo, Y.H.; Kim, S.B.; Hwang, B.Y.; Lee, M.K. Synthesis and biological evaluation of resveratrol derivatives as melanogenesis inhibitors. *Molecules* **2015**, *20*, 16933–16945. [CrossRef] [PubMed]
24. Nawaz, W.; Zhou, Z.; Deng, S.; Ma, X.; Ma, X.; Li, C.; Shu, X. Therapeutic versatility of resveratrol derivatives. *Nutrients* **2017**, *9*, 1188. [CrossRef] [PubMed]
25. Chalal, M.; Vervandier-Fasseur, D.; Meunier, P.; Cattey, H.; Hierso, J.C. Syntheses of polyfunctionalized resveratrol derivatives using Wittig and Heck protocols. *Tetrahedron* **2012**, *68*, 3899–3907. [CrossRef]
26. Chalal, M.; Klinguer, A.; Echairi, A.; Meunier, P.; Vervandier-Fasseur, D.; Adrian, M. Antimicrobial activity of resveratrol analogues. *Molecules* **2014**, *19*, 7679–7688. [CrossRef]
27. Chalal, M.; Delmas, D.; Meunier, P.; Latruffe, N.; Vervandier-Fasseur, D. Inhibition of cancer derived cell lines proliferation by newly synthesized hydroxylated stilbenes and ferrocenyl-stilbene analogs. Comparison with resveratrol. *Molecules* **2014**, *19*, 7850–7868. [CrossRef]
28. Latruffe, N.; Vervandier-Fasseur, D. Strategic syntheses of vine and wine resveratrol derivatives to explore their effects on cell functions and dysfunctions. *Diseases* **2018**, *6*, 110. [CrossRef]
29. Belluti, F.; Fontana, G.; Dal Bo, L.; Carenini, N.; Giommarelli, C.; Zunino, F. Design, synthesis and anticancer activities of stilbene-coumarin hybrid compounds: Identification of novel proapopoptic agents. *Bioorg. Med. Chem.* **2010**, *18*, 3543–3550. [CrossRef]
30. Conti, C.; Desideri, N. New 4H-chromene-4-one and 2H-chromene derivatives as anti-picornavirus capsid-binders. *Bioorg. Med. Chem.* **2010**, *18*, 6480–6488. [CrossRef]
31. Lima, L.M.; Barreiro, E.J. Bio-isosterism: A useful strategy for molecular modification and drug design. *Curr. Med. Chem.* **2005**, *12*, 23–49. [CrossRef] [PubMed]
32. Bae, S.J.; Ha, Y.M.; Kim, J.A.; Park, J.Y.; Ha, T.K.; Park, D.; Chun, P.; Park, N.H.; Moon, H.R.; Chung, H.Y. A novel synthesized tyrosinase inhibitor: (*E*)-2-((2,4-dihydrophenyl)diazinyl)phenyl-4-methylbenzenesulfonate as an azo-resveratrol analog. *Biosci. Biotechnol. Biochem.* **2013**, *77*, 65–72. [CrossRef] [PubMed]
33. Li, C.; Xu, X.; Wang, X.J.; Pan, Y. Imine resveratrol analogues: Molecular design, Nrf2 activation and SAR analysis. *PLoS ONE* **2014**, *9*, e101455. [CrossRef] [PubMed]

34. Mayhoub, A.S.; Marler, L.; Kondratyuk, T.P.; Park, E.J.; Pezzuto, J.M.; Cushman, M. Optimization of the aromatase inhibitory activities of pyridylthiazole analogues of resveratrol. *Bioorg. Med. Chem.* **2012**, *20*, 2427–2434. [CrossRef] [PubMed]
35. Bellina, F.; Guazzelli, N.; Lessi, M.; Manzini, C. Imidazole analogues of resveratrol: Synthesis and cancer cell growth evaluation. *Tetrahedron* **2015**, *71*, 2298–2305. [CrossRef]
36. Santos, A.C.; Pereira, I.; Pereira-Silva, M.; Ferreira, L.; Caldas, M.; Magalhaes, M.; Figueiras, A.; Ribeiro, A.J.; Veiga, F. Nanocariers for resveratrol delivery: Impact on stability and solubility concerns. *Trends Food Sci. Technol.* **2019**, *91*, 483–497. [CrossRef]
37. Moshawih, S.; Mydin, R.B.S.M.N.; Kalakotla, S.; Jarrar, Q.B. Potential applications of resveratrol in nanocarriers against cancer: Overview and future trends. *J. Drug Deliv. Sci. Technol.* **2019**, *53*, 101187. [CrossRef]
38. Intagliata, S.; Modica, M.N.; Santagati, L.M.; Montenegro, L. Strategies to improve resveratrol systemic and topical bioavailability: An update. *Antioxidants* **2019**, *8*, 244. [CrossRef]
39. Lian, B.; Wu, M.; Feng, Z.; Deng, Y.; Zhong, C.; Zhao, X. Folate-conjugated human serum albumin-encapsulated resveratrol nanoparticles: Preparation, characterization, bioavailability and targeting of liver tumors. *Artif. Cells Nanomed. Biotechnol.* **2019**, *47*, 154–165. [CrossRef]
40. Xiao, Y.; Chen, H.; Song, C.; Zeng, X.; Zheng, Q.; Zhang, Y.; Lei, X.; Zheng, X. Pharmacological activities and structure-modification of resveratrol analogues. *Pharmazie* **2015**, *70*, 765–771.
41. Biasutto, L.; Mattarei, A.; Azzolini, M.; La Spina, M.; Sassi, N.; Romio, M.; Paradisi, C.; Zoratti, M. Resveratrol derivatives as a pharmacological tool. *Ann. N. Y. Acad. Sci.* **2017**, *1403*, 27–37. [CrossRef] [PubMed]
42. Giacomini, E.; Rupiani, S.; Guidotti, L.; Recanatini, M.; Roberti, M. The use of stilbene scaffold in medicinal chemistry and Multi Target Drug design. *Curr. Med. Chem.* **2016**, *23*, 2439–2489. [CrossRef] [PubMed]
43. Solladié, G.; Pasturel-Jacopé, Y.; Maignan, J. A re-investigation of resveratrol synthesis by Perkin reaction. Application to the synthesis of aryl cinnamic acids. *Tetrahedron* **2003**, *59*, 3315–3321. [CrossRef]
44. Das, J.; Pany, S.; Majhi, A. Chemical modifications of resveratrol for improved protein kinase C alpha activity. *Bioorg. Med. Chem.* **2011**, *19*, 5321–5333. [CrossRef]
45. Tudose, A.; Maj, A.; Sauvage, X.; Delaude, L.; Demonceau, A.; Noels, A.F. Synthesis of stilbenoids via the Suzuki-Miyaura reaction catalyzed by palladium N-heterocyclic carbene complexes. *J. Mol. Cat. A Chem.* **2006**, *257*, 158–166. [CrossRef]
46. Siddiqui, A.; Dandawate, P.; Rub, R.; Padhye, S.; Aphale, S.; Moghe, A.; Jagyasi, A.; Swamy, K.V.; Singh, B.; Chatterjee, A.; et al. Novel aza-resveratrol analogs: Synthesis, characterization and anti-cancer activity against breast cancer cell lines. *Bioorg. Med. Chem. Lett.* **2013**, *23*, 635–640. [CrossRef]
47. Kim, S.; Ko, H.; Park, J.E.; Jung, S.; Lee, S.K.; Chun, Y.J. Design, synthesis and discovery of novel *trans*-stilbene analogues as potent and selective human cytochrome P450 1B1 inhibitors. *J. Med. Chem.* **2002**, *45*, 160–164. [CrossRef]
48. Kotora, P.; Sersen, F.; Filo, J.; Loos, D.; Gregan, J.; Gregan, F. The scavenging of DPPH, galvinoxyl and ABTS radicals by imine analogs of resveratrol. *Molecules* **2016**, *21*, 127. [CrossRef]
49. Song, Y.M.; Ha, Y.M.; Kim, J.A.; Chung, K.W.; Uehara, Y.; Lee, K.J.; Chun, P.; Byun, Y.; Chung, H.Y.; Moon, H.R. Synthesis of novel azo-resveratrol, azo-oxyresveratrol and their derivatives as potent tyrosinase inhibitors. *Bioorg. Med. Chem. Lett.* **2012**, *22*, 7451–7455. [CrossRef]
50. Tang, Y.Z.; Liu, Z.Q. Free-radical scavenging effect of carbazole derivatives on DPPH and ABTS radicals. *Cell Biochem. Funct.* **2007**, *2*, 149–158. [CrossRef]
51. Lu, J.; Li, C.; Chai, Y.F.; Yang, D.Y.; Sun, C.R. The anti-oxidant effect of imine resveratrol analogues. *Bioorg. Med. Chem. Lett.* **2012**, *22*, 5744–5747. [CrossRef] [PubMed]
52. Wright, J.S.; Johnson, E.J.; DiLabio, G.A. Predicting the activity of phenolic antioxidants: Theoretical method, analysis of substituents effects, and application to major families of antioxidants. *J. Am. Chem. Soc.* **2001**, *123*, 1173–1183. [CrossRef] [PubMed]
53. Zhang, Y.; Zou, B.; Pan, Y.; Liang, H.; Yi, X.; Wang, H. Antioxidant activities and transition metal ion chelating studies of some hydroxyl Schiff base derivatives. *Med. Chem. Res.* **2012**, *21*, 1341–1346. [CrossRef]
54. Anekonda, T.S. Resveratrol–A boon for treating Alzheimer's disease? *Brain Res. Rev.* **2006**, *52*, 316–326. [CrossRef] [PubMed]
55. Jang, J.H.; Surh, Y.J. Protective effect of resveratrol on β-amyloid-induced oxidative PC12 cell death. *Free Rad. Biol. Med.* **2003**, *34*, 1100–1110. [CrossRef]

56. Mancino, A.M.; Hindo, S.S.; Kochi, A.; Lim, M.H. Effects of clioquinol on metal-triggered amyloid-β aggregation revisited. *Inorg. Chem.* **2009**, *48*, 9596–9598. [CrossRef]
57. Martinez, A.; Alcendor, R.; Rahman, T.; Podgorny, M.; Sanogo, I.; McCurdy, R. Ionophoric polyphenols selectively bind Cu^{2+}, display potent antioxidant and anti-amyloidogenic properties, and are non-toxic toward *Tetrahymena thermophila*. *Bioorg. Med. Chem.* **2016**, *24*, 3657–3670. [CrossRef]
58. Yang, X.; Qiang, X.; Li, Y.; Luo, L.; Xu, R.; Zheng, Y.; Cao, Z.; Tan, Z.; Deng, Y. Pyridoxine-resveratrol hybrids Mannich base derivatives as novel dual inhibitors of AChE and MAO-B with anti-oxidant and metal-chelating properties for the treatment of Alzheimer's disease. *Bioorg. Chem.* **2017**, *71*, 305–314.
59. Battaini, G.; Monzani, E.; Casella, L.; Santagostini, L.; Pagliarin, R. Inhibition of the catecholase activity of biomimetic dinuclear copper complexes by kojic acid. *J. Biol. Inorg. Chem.* **2000**, *5*, 262–268. [CrossRef]
60. Lima, L.L.; Lima, R.M.; da Silva, A.F.; do Carmo, A.M.R.; da Silva, A.D.; Raposo, N.R.B. Azastilbene analogs as tyrosinase inhibitors: New molecules with depigmenting potential. *ScientificWorldJournal* **2013**, *2013*, 274643. [CrossRef]
61. Zimmermann-Franco, D.C.; Esteves, B.; Lacerda, L.M.; de Oliveira Souza, I.; dos Santos, J.A.; de Castro Campos Pinto, N.; Scio, E.; da Silva, A.D.; Macedo, G.C. In vitro and in vivo anti-inflammatory properties of imine resveratrol analogues. *Bioorg. Med. Chem.* **2018**, *26*, 4898–4906. [CrossRef] [PubMed]
62. Coimbra, E.S.; Santos, J.A.; Lima, L.L.; Machado, P.A.; Campos, D.L.; Pavan, F.R.; Silva, A.D. Synthesis, antitubercular and leishmanicidal evaluation of resveratrol analogues. *J. Braz. Chem. Soc.* **2016**, *27*, 2161–2166. [CrossRef]
63. Kucinska, M.; Piotrowska, H.; Luczak, M.W.; Mikula-Pietrasik, J.; Ksiazek, K.; Wozniak, M.; Wierzchowski, M.; Dudka, J.; Jäger, W.; Murias, M. Effects of hydroxylated resveratrol analogs on oxidative stress and cancer cells death in human acute T cell leukemia cell line: Prooxidative potential of hydroxylated resveratrol analogs. *Chem. Biol. Interact.* **2014**, *209*, 96–110. [CrossRef] [PubMed]
64. Cheng, L.X.; Tang, J.J.; Luo, H.; Jin, X.L.; Dai, F.; Yang, J.; Qian, Y.P.; Li, X.Z.; Zhou, B. Antioxidant and antiproliferative activities of hydroxyl-substituted Schiff bases. *Bioorg. Med. Chem. Lett.* **2010**, *20*, 2417–2420. [CrossRef]
65. Ronghe, A.; Chatterjee, A.; Singh, B.; Dandawate, P.; Murphy, L.; Bhat, N.K.; Padhye, S.; Bhat, H.K. Differential regulation of estrogen receptors α and β by 4-(E)-{(4-hydroxyphenylimino)-methylbenzene-1,2-diol}, a novel resveratrol analog. *J. Steroid Biochem. Mol. Biol.* **2014**, *144*, 500–512. [CrossRef]
66. Ronghe, A.; Chatterjee, A.; Singh, B.; Dandawate, P.; Abdalla, F.; Bhat, N.K.; Padhye, S.; Bhat, H.K. 4-(E)-{(p-tolylimino)-methylbenzene-1,2-diol}, a novel resveratrol differentially regulates estrogen receptors α and β in breast cancer cells. *Toxicol. Appl. Pharmacol.* **2016**, *301*, 1–13. [CrossRef]
67. Xu, H.; Zeng, X. Synthesis of diaryl-azo derivatives as potential antifungal agents. *Bioorg. Med. Chem. Lett.* **2010**, *20*, 4193–4195. [CrossRef]
68. Deri, B.; Kanteev, M.; Goldfeder, M.; Lecina, D.; Guallar, V.; Adir, N.; Fishman, A. The unravelling of the complex pattern of tyrosinase inhibition. *Sci. Rep.* **2016**, *6*, 34993. [CrossRef]
69. Algarni, A.S.; Hargreaves, A.J.; Dickenson, J.M. Activation of transglutaminase 2 by nerve growth factor in differentiating neuroblastoma cells: A role in cell survival and neurite outgrowth. *Eur. J. Pharmacol.* **2018**, *820*, 113–129. [CrossRef]

 © 2020 by the authors. Licensee MDPI, Basel, Switzerland. This article is an open access article distributed under the terms and conditions of the Creative Commons Attribution (CC BY) license (http://creativecommons.org/licenses/by/4.0/).

MDPI
St. Alban-Anlage 66
4052 Basel
Switzerland
Tel. +41 61 683 77 34
Fax +41 61 302 89 18
www.mdpi.com

Molecules Editorial Office
E-mail: molecules@mdpi.com
www.mdpi.com/journal/molecules